Zillions of Practice Problems

Advanced Algebra

Zillions of Practice Problems
Advanced Algebra

Stanley F. Schmidt, Ph.D.

Polka Dot Publishing

© 2019 Stanley F. Schmidt
All rights reserved.

ISBN: 978-1-937032-10-4

Library of Congress Catalog Number: 2013900926
Printed and bound in the United States of America

Polka Dot Publishing   Reno, Nevada

To order copies of books in the Life of Fred series,

visit our website **LifeofFred.com**

Questions or comments? Email the author at lifeoffred@yahoo.com

Third printing

Zillions of Practice Problems Advanced Algebra was illustrated by the author with additional clip art furnished under license from Nova Development Corporation, which holds the copyright to that art.

for Goodness' sake

or as J.S. Bach—who was never noted for his plain English—often expressed it:

Ad Majorem Dei Gloriam
(to the greater glory of God)

If you happen to spot an error that the author, the publisher, and the printer missed, please let us know with an email to: lifeoffred@yahoo.com

 As a reward, we'll email back to you a list of all the corrections that readers have reported.

What This Book Is All About

The first practice book, *Zillions of Practice Problems for Beginning Algebra*, was well received. It has served a definite need for many students of beginning algebra.

With that encouragement, I have spent months creating this practice book for advanced algebra.

In *Life of Fred: Advanced Algebra Expanded Edition*, there are *Your Turn to Play* sections after each topic. And each *Your Turn to Play* offers completely worked out solutions to each question.

At the end of each of the ten advanced algebra chapters are problem sets. In addition, there are five "Looking Back" chapters that review beginning algebra. The answers to all these questions are given in the book.

The book you are now holding, *Zillions of Practice Problems Advanced Algebra*, contains a massive number of new exercises—all keyed to *Life of Fred: Advanced Algebra Expanded Edition*.

As you work through each chapter in *Life of Fred: Advanced Algebra Expanded Edition*, you may do as many of the problems as you like in the corresponding chapter in this book.

Each of the problems is fully worked out. Some of the discussions are more than a page long.

There are nine examples of radical equations (like $\sqrt{x^2 - x + 16} = 6$). More than enough to learn how to solve them.

There are eleven completely worked out examples of linear programming like this problem: There were a lot of rocks in my backyard that I needed to remove. I had two options. I could either use a hammer and wheelbarrow or use dynamite and a truck.

With the hammer and wheelbarrow, my cost would be $1/day and I would get injured about 0.2 times each day. (On the average, I would get injured once every 5 days.)

With the dynamite and a truck, my cost would be $50/day and I would get injured about 0.1 times each day.

I had $817 to spend on the project and I could experience at most 5 injuries. (After five injuries I would quit and do something a lot less dangerous.)

I can remove 4 tons of rocks each day using hammer and wheelbarrow and 20 tons using dynamite and a truck. What is my best course of action in order to remove as much rock as possible?

Linear programming word problems look hard to many students. After you have done four or five of them, they start to look easy. After you have done eleven of them (in this book) and all the ones in *Life of Fred: Advanced Algebra Expanded Edition*, they become boringly easy. (That's our goal!)

Many other second-year algebra books (also called Advanced Algebra or Algebra 2) contain *zero* linear programming problems.

Some also skip:
- permutations
- matrices
- sigma notation
- math induction proofs
- partial fractions, and
- the change-of-base rule for logs

Both this book and *Life of Fred: Advanced Algebra Expanded Edition* cover all of these topics. They belong in any decent second-year algebra course.

Many other math publishers leave out these topics. The results are:

1) The students don't realize that the topics have been omitted.

2) The students finish the book more easily.

3) Everything is fine until they hit their SAT exams or later math courses. Then they find out the truth.

CHAPTERS IN THIS BOOK

Each advanced algebra chapter in this book is divided into two parts.

☆ The first part takes each topic and offers a zillion problems.

☆ The second part is called the 𝔐ixed 𝔅ag. It consists of a variety of problems from the chapter and review problems from the beginning of the book up to that point.

ELIMINATING TEMPTATION

The solutions and answers are all given in the back half of the book. The first question in this book is numbered "35." The second one is "101." In most ordinary books, they are numbered, "1, 2, 3 . . ." which is really silly when you think about it. In those books, when you look up the answer to "1" you might accidentally see the answer to "2" and that would spoil all the fun.

It usually takes three or four seconds (I timed it) to locate a solution in the back of the book.

YOUR FUTURE

This is the last bit of high school algebra that you will need. *Life of Fred: Advanced Algebra Expanded Edition* covers the material in 101 daily lessons. If you tuck in a lot of practice with this book, you might fluff things up to 140 or 160 days. Or, horrors, 180 days! But that is still less than half a year.

After advanced algebra, high school geometry, and a course in trigonometry, you have all the high school math required by virtually every major university.

With my best wishes for your future,
Stan Schmidt

Contents

(Chapters 1½, 2½, 3½, 4½, and 7½ are the five "Looking Back" chapters that review beginning algebra.)

Chapter 1
- Median Averages. 13
- Cross Multiplying. 13
- Constants of Proportionality. 14
- Using Constants of Proportionality. 14

Chapter 1½
- Exponents. 17
- Fractional Exponents and Square Roots. 17
- Rationalizing the Denominator. 18

Chapter 2
- Radical Equations. 19
- Imaginary Numbers. 19

Chapter 2½
- Venn Diagrams. 21
- Significant Digits. 22
- Scientific Notation. 22

Chapter 3
- The Meaning of Logarithms. 23
- The Laws of Logs. 23
- Change of Base Rule. 24
- Exponential Equations. 25

Chapter 3½
- The Language of Graphing. 27
- Graphing by Point-plotting. 27

Chapter 4
    Slope. 28
    Distance between Two Points. 29
    Slope-intercept Form of the Line. 29
    Double-intercept Form of the Line. 30
    Point-slope Form of the Line. 30
    Two-point Form of the Line.. 30

Chapter 4½
    Factoring—Common Factors.. 32
    Factoring—Easy Trinomials. 32
    Factoring—Grouping. 33
    Factoring—Harder Trinomials.. 33
    Simplifying Fractions.. 33
    Adding Fractions. 33
    Subtracting Fractions. 34
    Multiplying and Dividing Fractions.. 34
    Solving Fractional Equations.. 34
    Pure Quadratic Equations. 35
    Quadratic Formula. 35

Chapter 5
    Solving Systems of Equations by Elimination.. 36
    Solving Systems of Equations by Substitution. 36
    Solving Systems of Equations by Graphing.. 36
    Inconsistent and Dependent Systems of Equations. 37
    Determinants. 37
    Solving Systems of Equations by Cramer's Rule. 38

Chapter 6
    Ellipse. 40
    Circle. 43
    Parabola. 44
    Hyperbola.. 44
    Graphing Inequalities in Two Variables. 45

Chapter 7
- Definition of a Function........................... 47
- Range of a Function............................... 49
- Functions Represented as Ordered Pairs............. 49
- One-to-one Functions.............................. 50
- Functions That Are Onto........................... 51
- Inverse Functions................................. 51

Chapter 7½
- The Degree of a Polynomial........................ 54
- Long Division of Polynomials...................... 54

Chapter 8
- Partial Fractions................................. 55
- Linear Programming................................ 56
- Math Induction Proofs............................. 58

Chapter 9
- Arithmetic Sequences and Series................... 62
- Geometric Sequences and Series.................... 62
- Sigma Notation.................................... 63
- Matrices.. 65

Chapter 10
- The Fundamental Principle......................... 70
- Permutations...................................... 71
- Combinations...................................... 71
- The Binomial Formula and Pascal's Triangle........ 73

All the answers worked out in complete detail 78

Index 233

Chapter One

First part: Problems on Each Topic

Median Averages

35. Find the median average of 6, 8, 22, 49, 51
101. Find the median average of 57, 21, 21, 33, 999
146. Find the median average of 88, 23, 44, 90
201. Find the mean and the median averages of 7, 8, 10, 16, 14
273. Find a value for x so that the median average of 6, x, 10 is less than the mean average of those three numbers. [Not easy.]

Cross Multiplying

62. Solve $\dfrac{6}{2x} = \dfrac{12}{3x + 5}$

431. Solve $\dfrac{x + 1}{7} = \dfrac{5x - 3}{21}$

674. Solve $\dfrac{8}{x + 3} = \dfrac{4}{2x - 7}$

919. Solve $\dfrac{x - 9}{5} = \dfrac{-4}{x}$

1010. Solve $\dfrac{x - 2}{3} = \dfrac{-1}{x - 6}$

1028. Solve $\dfrac{-1}{5x^2 - 2x + 3} = \dfrac{3}{20x - 17}$

1261. Which of these can be solved by cross multiplication?
You are not being asked to solve them, just identify them.

First example: $\dfrac{x^2 + 6}{5x} = \dfrac{3.388x}{921}$ ☐ yes ☐ no

Second example: $\dfrac{3}{7x^2} = \dfrac{5}{x - 2} = \dfrac{x + 6}{8}$ ☐ yes ☐ no

Third example: $\dfrac{\pi x}{7} = \dfrac{4}{9}$ ☐ yes ☐ no

Fourth example: $\dfrac{5 + x}{3} = \dfrac{7x}{9} + 4$ ☐ yes ☐ no

Chapter One

Constants of Proportionality

Translate the English into an equation:

50. The cost (c) of a trip varies directly as the number of miles (m) driven.

104. The weight (w) of a rhino varies directly as the cube of its height (h).

485. The life of a car (L) as measured in months varies inversely as the average number of miles (m) that it is driven each day.

351. The time (t) it takes to plow a square field varies directly as the square of the length of one of the sides (s).

Twice the height
Eight times the weight

1064. In an evening of studying, the number of facts (F) learned varies directly as the square root of the number of hours (h) spent.

1102. The probability (p) of seeing something educational on television varies directly as the number of hours (h) you sit in front of the television and inversely as your intelligence (i).

Using Constants of Proportionality

There are three steps in doing these problems. ① Find the equation, which is what you did on the top half of this page. ② Use the given information to find the value of k in the equation. ③ Use the equation found in step one, the value of k found in step two, and values given in the problem to find the answer.

295. The cost (c) of cleaning up junk varies directly as the weight (w). If it cost $80 to clean up 30 pounds of junk, how much would it cost to clean up 50 pounds?

Does your backyard look like this?

Chapter One

311. The number of drops of sweat (s) varies directly as the number of miles (m) that you walk.

If you walk 12 miles, you will generate 5 drops of sweat.

How many drops of sweat is associated with walking 15 miles?

704. If a 4-foot Santa weighs 80 pounds, how much would a 6-foot Santa weigh? The weight of an object (w) varies as the cube of the height (h).

1225. How high (h) a fireworks rocket goes varies directly as the square root of the amount of powder (p) in the rocket.

A rocket with 38 pounds of powder will ascend 100 feet. How high will a rocket with 57 pounds of powder ascend?

Chapter One

Second part: the 𝓜ixed 𝓑ag: a variety of problems from this chapter

624. Translate into an equation: The noise (n) at a party varies directly as the square of the number of people (p) at the party.

1300. Sometimes it is very easy to turn an equation into a proportion and then use cross multiplying. For example, turn $\frac{a}{b} = c$ into $\frac{a}{b} = \frac{c}{1}$

Solve $\frac{3x + 1}{2x} = x$

884. The time (t) it takes to drive to my house from your house varies inversely as the speed (s) you are driving. At 50 miles per hour, it would take 8 hours to get there. How long would it take at 40 miles per hour?

16

Chapter One and a Half

Problems on Each Topic

Exponents

281. $x^3 x^5$
 $y^4 y^2 y^3$
 z^6 / z^3

644. $a^{10} a^{10}$
 $(b^7)^3$
 c^8 / c^4

904. d^{-4}/d^{-6}
 $(m^7 n^2)^3$
 a^0

1055. $(x^4 y^6)^4$
 z^{-3}/z^{-6}
 w^{-100}

401. $(8^x)^y$
 9^{-2}
 $1/a^{-7}$

83. $x^{0.4} x^{0.6}$
 $(y^w)^{3w}$
 $zzzzz^4$

Fractional Exponents and Square Roots

844. $y^{1/2}$
 $27^{1/3}$
 $\sqrt{2}\;\sqrt{8}$

1246. Simplify $\sqrt{18}$
 Simplify $\sqrt{x^5 y}$
 Simplify $1000^{1/3}$

107. $27^{2/3}$
 $64^{-1/2}$
 $(x^{1/2} y^{1/3})^{36}$

709. $(6x^2 + 48y^{-3x+z} - \sin x + \pi)^0$

17

Chapter 1½

Rationalizing the Denominator

1279. $\dfrac{8}{\sqrt{w}}$

303. $\dfrac{7}{\sqrt{x+6}}$

341. $\dfrac{\sqrt{x}+5}{\sqrt{x}}$

729. $\dfrac{\sqrt{w}}{\sqrt{xyz}}$

599. $\dfrac{7}{\sqrt{x}+5}$

1077. $\dfrac{43}{\sqrt{w}-7}$

1249. $\dfrac{x-y}{\sqrt{x-y}}$

1303. $\dfrac{x-y}{\sqrt{x}-\sqrt{y}}$

246. $\dfrac{a+\sqrt{c}}{a-\sqrt{c}}$

and just for fun . . .

491. $\dfrac{4}{3+x+\sqrt{y}}$

Chapter Two

First part: Problems on Each Topic

Radical Equations

A radical equation is an equation in which the unknown is under the radical sign.

283. $\sqrt{x+44} = 7$

1001. $8 + \sqrt{w+13} = 12$

1046. $\sqrt{70-y} - 2 = y$

1270. $\sqrt{x + 7x^2 + 3823} = -17$

71. $\sqrt{x^2 - x + 16} = 6$

168. $\sqrt{w+13} - 7 = w$

Imaginary Numbers

497. Simplify $\quad i^5$
$\qquad\qquad\qquad 3 + 4i + 7 - 6i$
$\qquad\qquad\qquad (2 + 5i)(8 + i)$

171. $\quad 5i + 7i - 8i$
$\qquad i^3$
$\qquad (4i)^2$

849. $\qquad\qquad\qquad i^{102}$
$\qquad\qquad\qquad (6 + 2i)^2$
$\qquad\qquad\qquad (-7i)(-8i)$

1019. Solve $x^2 = -1$

1031. Solve $y^2 + 5 = 3$

1122. $\dfrac{44 + 24i}{4}$

659. Place $\dfrac{2 + 3i}{5 + 6i}$ in the form a + bi where a and b are numbers.

Hint: When you had an expression like $\dfrac{2 + \sqrt{3}}{5 + \sqrt{6}}$ you multiplied top and bottom by the conjugate of the denominator.

251. Place $\dfrac{7}{4 - i}$ in the form a + bi.

Chapter 2

Second part: the 𝔐ixed 𝔅ag: a variety of problems from this chapter and previous material

467. Solve $\sqrt{22 - 3x} + 3 = 7$

819. Find the mean and median averages of 3.5, 4.3, 4.2

1222. Solve $\dfrac{2}{x - 8} = \dfrac{x - 1}{-3}$

371. When pirates attack a ship, the loot (L) each of them gets varies inversely as the number of pirates (p) doing the stealing. If 16 pirates attack the *Pinta*, they would each get 60 pieces of silver. If 12 pirates attack the *Pinta*, how many pieces of silver would each of them get?

    Juan Rodriguez Bermeo was the lookout on the *Pinta*. He hadn't seen land in five weeks. At about 2 a.m. on October 2, 1492, he spotted land in The Bahamas. He let everyone on the ship (and on the other two ships, the *Santa Maria* and the *Niña*) know that he had sighted land.

455. Simplify $\sqrt{50}$ Recall: $\sqrt{18} = \sqrt{9}\sqrt{2} = 3\sqrt{2}$

544. Solve $x^2 + 3x + 7 = 0$ Hint: Use the quadratic formula. You will get answers that involve i.

909. Place $\dfrac{-2 + i}{2 - i}$ in the form $a + bi$.

1192. $(3i)(4)(-5i)(2)$

313. $i^{1,000,000}$

285. $(7 + 3i)(2 - 3i)$

629. Simplify $\sqrt{27x^5}$

1282. $(-1 + \sqrt{3}\,i)^3$

Chapter Two and a Half

Problems on Each Topic

Venn Diagrams

89. Draw a Venn diagram of the set of all things that contain wood (W) and the set of all furniture (F).

377. Draw a Venn diagram of the set of all cats (C) and the set of all pizzas (P).

1285. Draw a Venn diagram of the set of all humans (H) and the set of all people who currently live in Paris (P).

437. Draw a Venn diagram of the intersection of the set of all weasels in the world (W) and all animals in America (A). Draw W ∩ A, which is the set of all things that are in both W and A.

1082. Draw a Venn diagram of the union of the set of all log cabins (L) and the set of all buildings (B).

1197. Draw a Venn diagram of the intersection of all log cabins (L) and the set of all buildings (B). L ∩ B means all those things that are both in L and in B.

594. I have 40 rabbits. I know that 18 of them like carrots, 25 like lettuce, and 10 like both carrots and lettuce. How many of them like neither?

1058. Joe owns 37 boats and ships.

    23 of them can go on the ocean ($\mathcal{O}$).

    20 have sails ($\mathcal{S}$).

    22 have a picture on Fred painted on the hull ($\mathcal{F}$).

    15 are ocean-going and have a picture of Fred on the hull.

    13 have sails and a picture of Fred on the hull.

    10 are ocean-going and have sails.

    7 are ocean-going, have sails, and have a picture of Fred.

How many are not ocean-going, have no sails, and no picture of Fred?

Chapter 2 ½

Significant Digits

151. You start counting significant digits with the first non-zero digit. Underline the first non-zero digit in:
        53.007
        3000
        00049.900
        60.6070

176. You stop counting significant digits with $\begin{cases} \text{the last non-zero digit or} \\ \text{the last gratuitous zero} \end{cases}$
        ... whichever occurs later.

A gratuitous zero is a zero that didn't have to be there, a zero that was voluntarily put there. (In restaurant talk, a gratuity is a tip.)

Double underline the gratuitous zeros in:
        5000
        30.70
        239.0809
        298.00

689. Put a "☞" under the last non-zero digit or the last gratuitous zero, whichever occurs later.
    9303.0   8000.   47   2020   300.00

934. Underline the first non-zero digit, put a "☞" under the last non-zero digit or the last gratuitous zero, whichever occurs later, and state how many significant digits are in each number.
        0.003004
        170
        2.0003
        39.50
        0005.006
        8,000,000

Scientific Notation

980. A number in scientific notation is in the form $d \times 10^n$ where $1 \leq d < 10$
    Put into scientific notation:
    600   0.005   30.08   2 10^6

854. Actually, a number in scientific notation is in the form $d \times 10^n$ where $1 \leq d < 10$ isn't quite true. Notice that −513.7 in scientific notation is -5.137×10^2. Rewrite the rule to take into account negative numbers. (This is hard because it involves English, which is often much more difficult than pure math.)

Chapter Three

First part: Problems on Each Topic

The Meaning of Logs (Logarithms)

$\log_a b$ means "a to what power will equal b?" $a^x = b$

92. Simplify:
    $\log_3 9$
    $\log_{10} 1000$
    $\log_{17} 17$

407. Simplify:
    $\log_5 \sqrt{5}$
    $\log_3 \frac{1}{3}$
    $\log_{82} 1$

1061. Simplify:
    $\log_{\sqrt{6}} 6$
    $\log_{0.1} 100$
    $\log_2 2^6$

The Laws of Logs

$\log mn = \log m + \log n$ The Product Rule
$\log m/n = \log m - \log n$ The Quotient Rule
$\log m^n = n \log m$ The Power Rule or The Birdie Rule

734. $\log 5x = ?$
      $\log \frac{2}{y} = ?$
      $\log (4 + w)(v) = ?$

1034. $\log x^7 = ?$
       $\log (w + \pi)^6 = ?$
       $\log \frac{x - 3y}{x - 3y} = ?$

1087. Which of these is nonsense?
      $\log_1 8$ $\log_0 298$ $\log_\pi 6$

Chapter 3

Change of Base Rule

The Change of Base Rule is $\quad\dfrac{\log_c a}{\log_c b} = \log_b a$

Multiplying both sides of
the equation by $\log_c b$ (and switching sides) $\quad (\log_c b)(\log_b a) = \log_c a$

     Both of these forms of the Change of Base Rule are given in various books. There is no need to memorize this rule. You probably won't use it that often. And when you do need it, it is sitting right here at the top of this page for you to use for the rest of your life.
     Spend your time learning how to use the formulas and procedures in Advanced Algebra, not memorizing all this stuff.
     Thought #1: If you memorize $\dfrac{\log_c a}{\log_c b} = \log_b a$ you will probably forget it by next week.

     Thought #2: If you have a job that requires knowing the Change of Base Rule, you will probably memorize it by the end of the second week just because you were using it a lot.

     Thought #3: This is similar to learning a language. The natural way to learn a language is by using it. Most five-year-olds in Greece can speak Greek, and they have never sat down and memorized Greek verbs.

     Thought #4: You probably don't know the word defenestration, but if you have ever been defenestrated, you would remember the meaning of that verb without any special effort on your part.

the act of throwing
out of a window

345. $\dfrac{\log_3 7}{\log_3 2}$

      $(\log_\pi 8)(\log_8 23)$

503. Simplify as much as possible $(\log_{10} 2)(\log_2 100)$

574. Convert $\log_5 9$ into an expression containing only base 10 logs.

739. Find an approximation for $\log_2 6$, rounding your answer to the hundredth place. Hints: Look at the previous problem. You will need a calculator that has a log key. The log key is used to find the $\log_{10}$ of a number. If you want to test your calculator, enter 55 and then hit the log key. Your answer will be approximately 1.74036268949424384553646107651 85. My calculator probably has more digits than yours.

Chapter 3

Exponential Equations

An exponential equation is an equation in which the unknown is in the exponent.

413. Solve $3^{x+2} = 7$

527. Solve $82^x = 85$

694. Solve $4 = 6^{x-7}$

1004. You put $287 into a savings account. It grows by 5% each year.
That means that after one year you would have $287(1.05)$.
After two years you would have $287(1.05)(1.05) = 287(1.05)^2$.
After three years, $287(1.05)^3$.
Approximately how many years would it take for you to have a million dollars in your account?
   Translation: $287(1.05)^x = 1{,}000{,}000$

1037. You have a million dollars. Each year the government taxes away 20% of what you own. *Approximately* how many years would it take for you to have $1000?
After one year you would have $1{,}000{,}000(0.8)$.
After x years you would have $1{,}000{,}000(0.8)^x$.

1107. You bought a new car. Each year it is only worth 85% of what is was in the previous year. To the nearest year approximate how long before it is worth only half of its original cost?
   $(0.85)^x = \frac{1}{2}$

859. During the summer, the number of weeds in my garden grows by 5% each day. How many days will it take for the number of weeds in my garden to triple?

Chapter 3

Second part: the 𝔐ixed 𝔅ag: a variety of problems from this chapter and previous material

604. $\log \dfrac{6y + 3}{2w - 1} = ?$ Use one of the laws of logs.

569. Solve $\dfrac{10x}{3x - 5} = \dfrac{1}{x + 3}$

261. Place $\dfrac{4 - 5i}{2 + 3i}$ in the form $a + bi$

824. Simplify $\sqrt{300}$

950. Solve $-5 = \sqrt{14w + 3}$

1112. Simplify $\log_{44} 44$
  $\log_8 64$
  $\log_{1/2} 2$

305. Every year the administration at KITTENS University changes their plans for the new Dean's Office.
  When the building was first designed, it was 1,200 square feet.
  Each year the plans increase its square footage by 7%.
  Approximately how many years before the Dean's Office would be 15,000 square feet?

Architect's drawing of the 15,000 square foot building

1007. Translate into an equation: The chance (C) that you will become a dean at KITTENS University varies directly as the square of the amount of money (m) that you donate to the President's Fund and inversely as the number of enemies (e) that you have made.

1273. Solve $6 = \sqrt{13x - x^2}$

1117. $\dfrac{\log_{11} 16}{\log_{11} 0.25}$

Chapter Three and a Half

Problems on Each Topic

The Language of Graphing

939. The point (6, 9) is in which quadrant?
      What is the abscissa of (6, 9)?
      If (a, b) is in quadrant II, what can you say about b?

983. What are the coordinates of the origin?
      Name a point (a, b) where a > 0 that does not lie in either quadrant I or quadrant IV.
      Is it possible for the graph of a straight line to lie in exactly two quadrants?

1147. If (a, b) lies in quadrant I, what can you say about a + b?
      If (a, b) lies in quadrant II, want can you say about ab?
      Give an example of a point with a positive ordinate that does not lie in either quadrant I or quadrant II.

Graphing by Point-plotting

112. Graph $y = 2x + 3$

271. Graph $y - 3x = -4$

297. Graph $x = 3y - 1$ Sometimes it is easier to name y values and then compute the corresponding x values. In this problem, if we let x equal 4, for example, we would have to work with fractions. On the other hand, if we let y be any integer (Integers = {...−3,−2,−1, 0, 1, 2, 3,...}), then the corresponding x values will also be integers.

461. Graph $2x + 4y = 8$

509. Graph $y = \ln x$ from x = 1 to x = 10
      You are being asked to graph the ln function, and you probably don't even know what that function is. This shows that you can graph any function by point-plotting as long as you can find the values of the function.
      If you have a calculator with a log key, it also probably has an ln key. If you punch in 2 and then hit the ln key, you will get something like 0.693 (or if you have a calculator like mine, you will find that the ln 2 ≈ 0.69314718055994530941723212145818. It doesn't matter. You just round it off and plot the point (2, 0.7). Plot a bunch of points from x = 1 to x = 10 until you see the shape of the curve and then connect the dots.

Chapter Four

First part: Problems on Each Topic

44. Find the slopes of lines ℓ, m, and n.

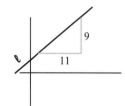
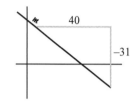
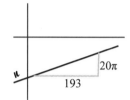

579. Given the coordinates of A are (3, 7). The coordinates of B are (15, 28).

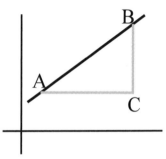

❶ Find the coordinates of C.
❷ Find the length of BC.
❸ Find the length of AC.
❹ Find the slope of AB.

864. Given the coordinates of A are (8, 13) and the coordinates of B are (11, 19), find the slope of the line through A and B. Hint: Use the same steps as in the previous problem.

944. Twice we have found the slope of a line given the coordinates of two of the points on the line. Now we will do it in general.
Given points (x_1, y_1) and (x_2, y_2) on line ℓ. Find the slope of ℓ.

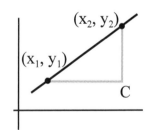

965. Given lines $ℓ_1$ and $ℓ_2$ are perpendicular, and they have slopes m_1 and m_2.

If $m_1 = \frac{2}{5}$ what does m_2 equal?

If $m_1 = 4$, what does m_2 equal?

If $m_1 = -0.01$ what does m_2 equal?

Chapter 4

Distance between Two Points

299. Given the coordinates of A are (2, 6).
The coordinates of B are (8, 14).

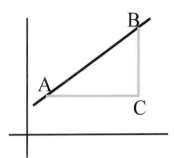

❶ Find the coordinates of C.
❷ Find the length of BC.
❸ Find the length of AC.
❹ Find the length of AB.

If this looks slightly familiar, it should. Steps ❶, ❷, and ❸ are identical to the ones in problem 579 on the previous page. We are starting at the same place (given two points) but are heading toward a different result. Instead of getting the slope between those two points, we are going to find the distance between them.

339. We have found the length of a segment given the coordinates of two endpoints on the line. Now we will do it in general.

Given points (x_1, y_1) and (x_2, y_2) on line ℓ. Find the length of the segment that joins those two points.

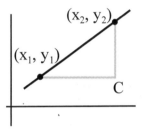

549. What is the distance between (7, 3) and (5, 8)?

664. What is the distance between (4, –2) and (–9, 1)?

759. What is the perimeter of the triangle with vertices (2, 4), (4, 6), and (5, –7)?

Slope-intercept Form of the Line y = mx + b

869. What is the equation of the line with a slope of ⅔ that intercepts the y-axis at (0, 7)?

924. What is the point at which y = 8.3x + 2.705 intercepts the y-axis?

1040. Without point-plotting, graph $y = \frac{2}{5}x + 4$

1137. Graph these six lines:

$y = \frac{2}{3}x + 4$ $\qquad$ $y = \frac{1}{4}x + 6$ $\qquad$ $y = \frac{-5}{3}x + 2$

$y = 6x + 1$ $\qquad$ $y = -3x + 7$ $\qquad$ $y = x + 5$

Chapter 4

Double-intercept Form of the Line $x/a + y/b = 1$

383. At what point does $\dfrac{x}{5} + \dfrac{y}{6} = 1$ intercept the x-axis?

719. Graphing lines in double-intercept form is even faster than graphing lines in y = mx + b form.

Graph these six lines:

$\dfrac{x}{2} + \dfrac{y}{4} = 1$ $\dfrac{x}{1} + \dfrac{y}{5} = 1$ $\dfrac{x}{7} + \dfrac{y}{3} = 1$

$\dfrac{x}{-3} + \dfrac{y}{-1} = 1$ $\dfrac{x}{6} + \dfrac{y}{\pi} = 1$ $\dfrac{x}{0.7} + \dfrac{y}{5} = 1$

Point-slope Form of the Line $m = \dfrac{y - y_1}{x - x_1}$
This line passes through the point (x_1, y_1) with a slope of m.

764. What is the equation of the line with a slope of 5 that passes through the point (4, −17)?

1276. Graph $\dfrac{2}{5} = \dfrac{y - 6}{x - 4}$

1291. What is the equation of the line that is perpendicular to the line through (2, 5) and (4, 8) and that passes through (8, 9)?

Two-point Form of the Line $\dfrac{y - y_1}{x - x_1} = \dfrac{y_2 - y_1}{x_2 - x_1}$
This line passes through the points (x_1, y_1) and (x_2, y_2).

289. What is the equation of the line that passes through (8, 9) and (2, 5)?

343. Find the equation of the dashed line.
The dashed line contains one of the diagonals of the rectangle.

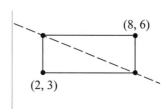

Chapter 4

Second part: the 𝕸ixed 𝕭ag: a variety of problems from this chapter and previous material

953. Combine log 67 + log xyz   Use one of the laws of logs.

321. Solve $x = \sqrt{6-x}$

609. Suppose you owned 1,000 flamingos. Suppose their population was decreasing by 8% each year because of the neighborhood cats that were attacking them.

    How long (to the nearest year) would it be before you had only 700 flamingos?

714. If the product of the abscissa and the ordinate of a point is positive, what quadrant or quadrants must it lie in?

754. Graph by point-plotting $y = \tan x$ from $x = 0$ to $x = 60$.

    You probably have not seen the tangent function (tan x) before. It is usually first encountered in trig. The tan key is on your calculator. If you punch in 20 and hit the tan key, you will get 0.3639702, which rounds to 0.4. Therefore, the point (20, 0.4) will be on your graph.

    The numbers on the x-axis will be a lot larger than those on the y-axis. One unit on the x-axis might be equal to 10. One unit on the y-axis might be equal to 0.2.

889. Graph $y = \dfrac{-2}{3}x + 8$

947. What is the equation of the line that passes through (–4, 7) and (3, 10)?

1043. Find the mean and median averages of 3, 9, 5, 6, 12

1067. Graph $\dfrac{x}{8} + \dfrac{y}{3} = 1$

1127. What is the distance between (–3, 5) and (3, –5)?

1152. What is the equation of the line with a slope of $\sqrt{2}$ that passes through the point (33, 55)?

Chapter Four and a Half

Problems on Each Topic

Factoring—Common Factors

293. Factoring is the opposite of multiplying. Multiplying is easier. Let us warm up by doing some multiplication.
$$4x(3x + 5y) = ?$$
$$-3y(6y^2 - 2y) = ?$$
$$7x^2y(3xy + 8x^4) = ?$$

554. Factor $\quad 40x^4y^6 - 50x^4y^7$

1306. Factor $\quad 18w^3 + 27w^4$
$\qquad\qquad 3x^9 - 12x^6y$
$\qquad\qquad 15xyz + 20x$

1342. Factor
$\qquad\qquad 9x^4 + 12x^3y + 16x^6y^6$
$\qquad\qquad 8w^5 + 16w^{10}$
$\qquad\qquad 100 + 30z$
$\qquad\qquad 49y^4 + 36z^4$

Factoring—Easy Trinomials (of the form $x^2 + bx + c$)

52. Factor $\quad x^2 + 9x + 20$

296. Factor
$\qquad\qquad x^2 + 20x + 100$
$\qquad\qquad y^2 + 11y + 18$
$\qquad\qquad z^2 + 6z + 9$
$\qquad\qquad 3x^2 + 27x + 54 \qquad$ Always look for a common factor first.

956. Factor $\quad y^2 - 12y + 11$
$\qquad\qquad w^2 - 2w - 15$
$\qquad\qquad x^2 + 3x + 40$
$\qquad\qquad 5z^2 + 45z + 70$

1142. Factor
$\qquad\qquad 7x^2 + 21x + 14$
$\qquad\qquad y^2 - 2y - 24$
$\qquad\qquad x^2 + 27xy + 50y^2$
$\qquad\qquad w^2 + 5w - 50$

Chapter 4 ½

Factoring—Grouping

389. Factor $x^3 + 7x^2 + 3x + 21$

500. Factor $3x^2y - xyw + 3xw - w^2$
$45x + 15xy + 6y + 2y^2$
$6ac + 12ad + c + 2d$
$8xy + 20x + 6y + 15$

Factoring—Harder Trinomials (of the form $ax^2 + bx + c$ where $a \neq 1$)

325. Factor $7x^2 + 37x + 10$
$2x^2 + 11x + 12$

634. Factor $12y^2 + y - 6$
$14x^2 + 35x + 14$
$4w^2 - 23w + 15$

929. Factor $3x^2 - 13x + 12$
$9y^2 + 21y + 10$
$40x^2 - 40x + 10$

Simplifying Fractions

The three steps are: factor the top, factor the bottom, cancel like factors.

95. $\dfrac{6x^2 + 7x - 3}{18x - 6}$

307. $\dfrac{x^2 - y^2}{3x^2 + 5xy + 2y^2}$

699. $\dfrac{49a^2 - c^2}{7a^2 + 8ac + c^2}$

Adding Fractions

Factor denominators, decorate top and bottom of each fraction until all denominators are alike, add.

206. $\dfrac{5}{x + 2} + \dfrac{3}{x - 7}$

Chapter 4 ½

277. $\dfrac{1}{x+5} + \dfrac{2x+11}{x^2+9x+20}$

315. $\dfrac{5}{x-6} + \dfrac{-3x-18}{x^2-36}$

986. $\dfrac{2}{x-5} + \dfrac{8}{5-x}$

Subtracting Fractions

894. $\dfrac{2x^2+7x-23}{x^2+11x+30} - \dfrac{x-1}{x+6}$

1072. $20 - \dfrac{x+6}{x+3}$

Multiplying and Dividing Fractions

To multiply fractions: top times top and bottom times bottom

1401. $\dfrac{7x+21}{3x^2-17x+10} \cdot \dfrac{x^2-25}{x^2+x-6}$

1450. $\dfrac{10x^2-30x-100}{4x^2-21x+5} \cdot \dfrac{4x^2-9x+2}{15x^2-60}$

To divide fractions: turn $\dfrac{a}{b} \div \dfrac{c}{d}$ into $\dfrac{a}{b} \cdot \dfrac{d}{c}$

1506. $\dfrac{16w^2-16w+3}{4w^2-17w+15} \div \dfrac{(4w-1)^6}{16w-20}$

Solving Fractional Equations

After factoring each denominator, multiply each term by an expression that all the denominators divide into evenly. If you multiply by something containing a variable, you must check your answer to eliminate any extraneous roots.

559. $\dfrac{3}{2x} + \dfrac{2}{x} = \dfrac{7}{10}$

829. $z + \dfrac{1}{z+2} = \dfrac{6z+1}{6}$

959. $\dfrac{4}{6-x} + \dfrac{2}{3x} = \dfrac{4}{3}$

1013. $\dfrac{x+10}{4x-24} + \dfrac{4}{x^2-8x+12} = \dfrac{5}{x-6}$

Chapter 4 ½

Pure Quadratic Equations

317. $x^2 = 49$
$y^2 = 10^6$
$z^2 = 33$

789. $6x^2 = 54$
$y^2 - 27 = 0$
$z^2 + 6z = 2(3z + 50)$

968. Find the length of the hypotenuse.

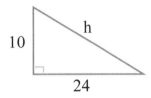

Quadratic Formula

General quadratic equations, $ax^2 + bx + c = 0$, may be solved by $x = \dfrac{-b \pm \sqrt{b^2 - 4ac}}{2a}$

161. Solve $3x^2 + 5x + 1 = 0$

899. Solve $4x^2 + 12x + 5 = 0$
$6x^2 + x + 2 = 0$

989. Solve $4x - 7 = -2x^2$
$(5x + 3)(2x - 5) = -24$

1345. Solve $(4x^2 - 3)(3x + 2) = (6x^2 + x)(2x - 5)$

639. I saw an advertisement on the Internet for farmland.

Opportunity of a Lifetime!
Giant rectangle of 7,500 square miles of high-quality farmland in Nevada. Length is 9 miles more than the width. Only $285.

I bought it even without looking at it. Rounding to the nearest mile, what are the dimensions of the rectangle?

Chapter Five

First part: Problems on Each Topic

Solving Systems of Equations by Elimination

256. Solve $\begin{cases} 5x + 2y = 41 \\ 3x - 2y = 15 \end{cases}$

279. Solve $\begin{cases} 7x + 3y = 33 \\ 9x - 6y = 3 \end{cases}$

992. Solve $\begin{cases} 3x + 5y = 41 \\ 2x - 6y = -38 \end{cases}$

1309. If I buy 3 dogs and 2 bags of dog food, it will cost $70. If I buy 4 dogs and 3 bags of dog food, it will cost $97. Let x = number of dogs and y = number of bags of dogfood.

How much does one dog cost? How much does a bag of dog food cost?

Solving Systems of Equations by Substitution

679. Solve using substitution $\begin{cases} 3x + y = 5 \\ 7x + 6y = -3 \end{cases}$

744. Solve $\begin{cases} 11x - 2y = 10 \\ x + 7y = 44 \end{cases}$ Hint: Solve the second equation for x and substitute into the first equation.

1157. Solve $\begin{cases} 3x - 5y = 7 \\ 2x - y = 7 \end{cases}$

Solving Systems of Equations by Graphing

226. Solve by graphing $\begin{cases} 3x + y = 5 \\ 7x + 6y = -3 \end{cases}$

700. Solve by graphing $\begin{cases} 4x + y = 5 \\ -3x + 2y = 0 \end{cases}$

Chapter 5

Inconsistent and Dependent Systems of Equations

794. There are two ways that you can tell if a pair of equations is inconsistent:
    A) What do their graphs look like?
    B) What do you get when you try to solve them by the elimination method?

971. There are two ways that you can tell if a pair of equations is dependent:
    A) What do their graphs look like?
    B) What do you get when you try to solve them by the elimination method?

1359. A system of equations that has exactly one solution (it isn't inconsistent and it isn't dependent) is called independent.

Classify each of these systems as   ❀ independent
                                    ❀ inconsistent or
                                    ❀ dependent

$$\begin{cases} 3x + 7y = 30 \\ 5x - 7y = 539 \end{cases}$$

$$\begin{cases} -2x + 4y = 44 \\ 2x + 6y = 100 \end{cases}$$

$$\begin{cases} 6x - y = 13 \\ -12x + 2y = 20 \end{cases}$$

$$\begin{cases} 4x + 5y = 8 \\ -12x - 15y = -24 \end{cases}$$

Determinants

333. Evaluate $\begin{vmatrix} 3 & 5 \\ 7 & 4 \end{vmatrix}$

539. Evaluate $\begin{vmatrix} 6 & 8 \\ -1 & 3 \end{vmatrix}$

            and $\begin{vmatrix} 9 & 2 \\ 0 & -3 \end{vmatrix}$

Chapter 5

769. What is the minor of 7 in the determinant $\begin{vmatrix} 6 & 7 & 9 \\ 2 & 5 & 1 \\ 0 & 3 & 5 \end{vmatrix}$

874. Evaluate $\begin{vmatrix} 4 & 6 & -7 \\ 0 & 3 & 5 \\ -2 & 2 & 8 \end{vmatrix}$

1202. Evaluate $\begin{vmatrix} 1 & 0 & 30 \\ 5 & 9 & -3 \\ 4 & 2 & -8 \end{vmatrix}$

1237. Evaluate $\begin{vmatrix} 6 & 2 & 7 & 1 \\ 3 & 0 & 0 & 6 \\ 9 & 1 & 0 & 4 \\ 9 & 2 & 0 & 3 \end{vmatrix}$

Solving Systems of Equations by Cramer's Rule

166. Solve for y using Cramer's rule. You do not have to evaluate the determinants.

$$\begin{cases} 6x + 7y - 3z = 44 \\ 2x - 4y + 8z = 39 \\ 5x + 6y + 9z = 82 \end{cases}$$

335. Solve for z. You do not have to evaluate the determinants.

$$\begin{cases} 2x + y + 8z = 9 \\ 7x + 3y - 5z = 8 \\ 3x - 8y + z = 7 \end{cases}$$

684. Solve for y. You do not have to evaluate the determinants.

$$\begin{cases} 3w + 5x + 8y + 2z = 1 \\ 2w + 7x + 3y - 3z = 3 \\ 5w - 3x - 9y + 8z = 4 \\ 7w + 4x + 2y + 5z = -3 \end{cases}$$

Chapter 5

Second part: the 𝔐ixed 𝔅ag: a variety of problems from this chapter and previous material

38. What is the least and the greatest number of quadrants that a straight line might lie in?

80. The pull (P) that I feel from my mother's carrot cake varies inversely with the distance (d) that I am from it. If I am 10 feet away from it, I feel a pull of 96 pounds drawing me toward it.
   When I'm only 4 feet away, I can see it. I can smell it. I can almost reach it. What is the force of attraction at that distance?

96. Evaluate $\begin{vmatrix} 5 & 2 & -1 \\ 3 & -3 & 0 \\ 4 & -1 & 0 \end{vmatrix}$

121. Simplify $\sqrt{6}\sqrt{12}$

181. Simplify $\log_{44} 44^{0.73}$
   $\log_{6.0735} 1$
   $\log 1000$

216. Solve $(2x + 5)(3x - 1) + 5 = 0$

266. Every year my book collection increases by 6%. In how many years (to the nearest year) will it take to double my collection?

287. What is the equation of the line with a slope of $\frac{3}{4}$ that passes through the point $(-5, -1)$?

784. Factor $50x^2 - 200y^2$
   $x^2 + 8x + 16$
   $15x^2 - 2x - 8$

1049. Solve for z using Cramer's rule. You do not have to evaluate the determinants.

$$\begin{cases} 3x + 2y - 35z = 2 \\ 9x - 8y + 67z = 3 \\ 2x + 3y + 4z = 4 \end{cases}$$

Chapter Six

First part: Problems on Each Topic

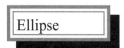

1177. The tops of cans of tuna are round (circles). If you look at one from an angle, the top becomes an ellipse.

The pan I fry my eggs in is circular. But if I add some milk and look at it from an angle, it looks like an ellipse.

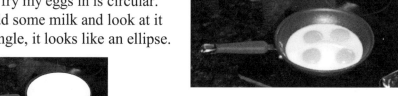

Your question: When kids draw pictures of the sun, those pictures are circles and not ellipses. When people in rocket ships photograph the sun, it is a circle and not an ellipse. Why doesn't it turn into an ellipse?

540. What is the equation of this ellipse?

834. Place $4x^2 + 25y^2 = 100$ into the standard form for ellipses, which is $\frac{x^2}{a^2} + \frac{y^2}{b^2} = 1$. Then graph the ellipse.

914. Graph $36x^2 + 16y^2 = 576$. What are the coordinates of the vertices?

1016. Graph $16(x - 3)^2 + (y - 5)^2 = 16$
What are the coordinates of the center and of the vertices?

1207. Complete the square for each of these:
$x^2 + 6x$ $y^2 - 10y$ $5z^2 + 40z$ $9w^2 + 36w$
     You will need to complete the square in order to put ellipses whose centers are not at the origin into standard form.

Chapter 6

1255. Place into standard form for an ellipse: $5x^2 + y^2 + 6y = 16$
You will need to turn $y^2 + 6y$ into $(y + ?)^2$ by completing the square.

1288. Place into standard form: $4x^2 - 16x + 9y^2 - 90y = -205$

186. Suppose you are given the equation of an ellipse in the form
$$ax^2 + bx + cy^2 + dy = e$$
where a, b, c, d, and e are numbers.

If $a = 0$ or $c = 0$ or $e = 0$, then we do not have an ellipse.

What does $b \neq 0$ or $d \neq 0$ tell you?

Hint: Look at the previous two problems.

small essay
Preview of Coming Attractions in Conics

Here, in a nutshell, is what is ahead for you in conics. The conics are the ellipse, the circle, the parabola, and the hyperbola.

The general equation for all of these conics is:
$$ax^2 + bxy + cy^2 + dx + ey + f = 0$$

Some secrets:

If $b = 0$, then we know . . .

   i) If a and c are both positive or both negative and $a \neq c$, then it is an ellipse. (Another way to say that is: $ac > 0$.)

   ii) If $a \neq 0$ and $a = c$, then it is a circle.

   iii) If a and c are of opposite sign, then it is a hyperbola. (Another way to say that is: $ac < 0$.)

   iv) If exactly one of a and c are nonzero, then it is a parabola.

If $b \neq 0$, we might have something like $6x^2 + 7xy + 2y^2 - 5x + 3y = 8$.

Then you have to wait until the fifteenth chapter of *Life of Fred: Calculus*!

Why? Because when we have an xy-term that is not zero, we have a conic that has been *rotated*.

Chapter 6

If it were an ellipse it might look like this: All we need to do is get rid of the bxy term and that will "un-rotate" the conic.

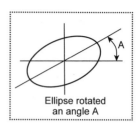
Ellipse rotated an angle A

If I told you all the steps necessary to eliminate that bxy term, that would make this work in advanced algebra look like child's play.
And if I did tell you, you would probably have nightmares tonight.

I, your reader, demand that you tell me!

But, it's long—like following a recipe to make Paupiettes. You may choose either nightmare . . .

OR

Eliminating the bxy term from $ax^2 + bxy + cy^2 + dx + ey + f = 0$	Paupiettes Fish in a Mold
Step 1: Draw a right triangle where the right leg is equal to b and the bottom leg is equal to a − c. 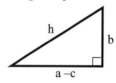	**Step 1:** Go fishing and catch a Dover sole or buy it in the store. Skin it and cut two thin fillets. (fill-LAYs)
Step 2: Find the length of the hypotenuse (h) using the Pythagorean theorem. $h = \sqrt{(a-c)^2 + b^2}$	**Step 2:** Butter a mold and stick the fillets in it making an X. Set it aside for a moment.
Step 3: Let $L = \dfrac{a-c}{h}$ Let $M = \sqrt{(1-L)/2}$ Let $N = \sqrt{(1+L)/2}$	**Step 3:** Make a fish quenelles by first finely chopping a bunch of pike or shrimp and then sticking it in a bowl that is inside another bowl filled with ice. Use a wooden spoon and smash it down into a paste.
Step 4: Everywhere you see an x in $ax^2 + bxy + cy^2 + dx + ey + f = 0$ substitute **Nx' − My'** and everywhere you see a y substitute	**Step 4:** Add some egg whites and some nutmeg, salt, and a little hot pepper sauce. Stir it into a thick paste.

Mx' + Ny' where x' and y' are new variables.

After you simplify all the algebra, the new equation in x' and y' will not have any xy term in it.

Then you can immediately identify what kind of conic you have using the Some secrets listed two pages ago.

Step 5: Very slowly add some chilled whipping cream while stirring with your wooden spoon. Stop when the quenelles get to the consistency of whipped cream.

Step 6: Fill the bowl from step 2 with the quenelles and leave enough room to fold the top ends of the fillets over the quenelles. That makes it look pretty.

Step 7: Stick the mold in a pan of hot water and bake the whole thing about a half hour.

Step 8: Serve it with an anchovy sauce (three smashed anchovy fillets in a white sauce).

Cooking is harder than calculus.

end of small essay

347. $\frac{(x-7)^2}{81} + \frac{(y+2)^2}{100} = 1$ is an ellipse centered at (7, –2) with a semi-major axis length of 10 and a semi-minor axis length of 9. How would $\frac{(x-7)^2}{100} + \frac{(y+2)^2}{100} = 1$ be described?

425. Put into the standard form for a circle $(x-h)^2 + (y-k)^2 = r^2$ and graph:
$16x^2 + 32x + 16y^2 - 96y = 9$

770. In English, tell how you would determine whether the point (7, 13) lies inside the circle $(x-3)^2 + (y-5)^2 = 81$.

993. Continuing the previous problem, determine whether the point (7, 13) lies inside the circle $(x-3)^2 + (y-5)^2 = 81$.

Chapter 6

Parabola (per-RA-beh-lah)

84. For all the other conics (ellipses, circles, and hyperbolas) their equations contain both x-squared and y-squared terms.
       Parabolas look like $y = ax^2 + bx + c$
        or they look like $x = ay^2 + by + c$,
       where a, b, and c are numbers.
$y = 3x^2$ points upward.
$y = -3x^2$ points downward.

Perhaps the easiest way to graph a parabola is by point-plotting.
    Graph $y = 2x^2 + 5$ from $x = -2$ to $x = 2$.

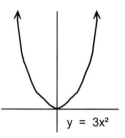

131. Graph $y = -4x^2 + 7x$ from $x = -5$ to $x = 5$. Generally, in point-plotting parabolas, you will need around five points to get a feeling for the shape of the curve. If you were to plot all eleven points from $x = -5$ to $x = 5$, you might be overdoing it. Or you might be someone who is paid by the hour instead of by the job.

470. Graph $y = x^2 - 4x + 12$ from $x = 1$ to $x = 3$. Please plot at least five points.

799. Redoing the previous graph with a "slight" change. Graph $y = x^2 - 4x + 12$ from $x = -8$ to $x = 12$. Please plot at least five points.

915. We will graph a parabola that opens to the right. Graph $x = 8y^2$ from $y = -3$ to $y = 3$.

Hyperbola (high-PER-beh-lah)

Hyperbolas and ellipses are brother and sister to each other.

An ellipse in standard form: $\dfrac{(x-h)^2}{a^2} + \dfrac{(y-k)^2}{b^2} = 1$

The corresponding hyperbola: $\dfrac{(x-h)^2}{a^2} - \dfrac{(y-k)^2}{b^2} = 1$

Both of these are centered at (h, k).
To graph the hyperbola, you first graph the ellipse.
All the steps are listed on the next page.

Chapter 6

How to Graph a Hyperbola

The Steps	How to Remember the Steps
1. Graph the ellipse.	1. Draw the egg.
2. Enclose the ellipse in a rectangle.	2. Box the egg.
3. Draw lines through the opposite corners of the box. (Another way of saying that is draw the diagonals of the rectangle and extend them into lines.)	3. X out the box.
4. Use those lines as asymptotes and draw the hyperbola. It will just touch the rectangle. The vertices of the hyperbola are the points where the hyperbola touches the rectangle.	4. Draw the hyperbola.

724. Graph $\dfrac{(x-2)^2}{9} - \dfrac{(y-4)^2}{25} = 1$

804. Graph $4x^2 - 25y^2 = 100$

960. Graph $-4x^2 + 25y^2 = 100$

You might notice that this equation is very much like the equation in the previous problem. Hyperbolas in which the x^2 term is positive open up in the x direction (horizontally). Hyperbolas in which the y^2 term is positive open up in the y direction (vertically).

1052. What are the coordinates of the center and of the vertices of the hyperbola $\dfrac{(x-7)^2}{100} - \dfrac{(y+23)^2}{144} = 1$

1162. What are the coordinates of the center and of the vertices of the hyperbola $-\dfrac{(x-7)^2}{100} + \dfrac{(y+23)^2}{144} = 1$

This problem is very similar to the previous problem.

Graphing Inequalities in Two Variables

395. Graph $y > \dfrac{3}{4}x + 2$

614. Graph $\dfrac{x^2}{7} + \dfrac{y^2}{9} \geq 1$

809. Graph $y < 2x^2 + 3$

875. Graph $\dfrac{(x-5)^2}{9} - \dfrac{(y+2)^2}{3} \leq 1$

Chapter 6

Second part: the 𝔐ixed 𝔅ag: a variety of problems from this chapter and previous material

65. $(3i)^3$

110. $\log_8 9 - \log_8 3 = ?$

309. Solve $(2x^2 + 3)(3x - 1) = 6x(x^2 + 3)$

331.
  Graph by point-plotting $y = x!$ from $x = 0$ to $x = 5$.
       This is called the factorial function. We will get to it in Chapter 10 of the Advanced Algebra book. The factorial function is probably on your calculator. Punch in 5 and then hit the ! key. The answer should be 120. The point (5, 120) indicates that the scale on the y-axis will be different than the one on the x-axis. The curve climbs very quickly.

443. ★ Find the center of the ellipse whose vertices are (5, 8) and (5, 12).
      ★ Find the length of the semi-major axis.
      ★ Can you find the length of the semi-minor axis?

1172. Solve $\dfrac{2x}{4-x} = \dfrac{3+x}{5x}$

1252. $(\log_8 9)(\log_9 64) = ?$

1294. Find the center and vertices of $100x^2 + 800x + 9y^2 + 90y = -925$

Chapter Seven

First part: Problems on Each Topic

You start with two sets. Call the first set the domain and the second set the codomain. A function is any rule that assigns to each member of the domain exactly one member of the codomain.

This definition takes a while for many students to understand. Maybe it's because it involves understanding English.

The math isn't hard. All you have to do is be able to count up to the number 1.

Look at the first member of the domain. Does it have exactly one assignment in the codomain?

Look at the second member of the domain. Does it have exactly one assignment in the codomain?

Look at each member of the domain. If each one has exactly one image in the codomain, you have a function.

If some member of the domain has no image in the codomain, it isn't a function.

If some member of the domain has two assignments in the codomain, it isn't a function.

41. Which of these are functions?

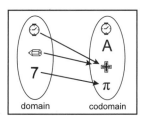 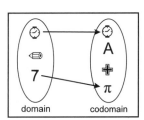 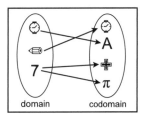

example 1   example 2   example 3

240. Sometimes, instead of drawing ellipses around the domain and the codomain, we just write:

> Nebraska → moon
> Kansas → purple
> South Dakota → #
> New Jersey → 8
> Kansas → coat

The items to the left of the → are in the domain, and the items to the right of the → are in the codomain.

Is this a function?

Chapter 7

Sometimes we name a function we have created. Probably the most popular name is f. It sounds nice to say, "function f."

Putting the name of the function over the arrows is one way to name the function.

337. Is f a function?   cuff $\xrightarrow{f}$ ❊

sleeve $\xrightarrow{f}$ rock

shoe $\xrightarrow{f}$ ❊

479. Is g a function?

g(rice) = can
g(mouse) = can
g(Harvard) = can
g(clock) = can
g(shovel) = rooster

In the previous two chapters, we had lots of numbers. Recall what it would be like to evaluate

$$\begin{vmatrix} 5 & 7 & 9 & 2 & 8 \\ 3 & 2 & 7 & 5 & 5 \\ 9 & 4 & 5 & 3 & 1 \\ 3 & 7 & 3 & 1 & 9 \\ 2 & 8 & 2 & 2 & 2 \end{vmatrix}$$

This sends chills up my spine. A determinant, with five rows and columns when evaluated by minors, turns into five 4x4 determinants.
Each 4x4 determinant would turn into four 3x3 determinants.
Each 3x3 determinant would become three 2x2 determinants.

Happily, this chapter on functions has very little computation. The only thing you have to do is be able to count up to one to decide whether a particular rule is a function.

Some parts of mathematics involve a lot of computation. Solving a radical equation like $\sqrt{x^2 + 33x} - \sqrt{2x - 1} = 4.27 + \sqrt{98x^2 - 21}$ involves zillions of lines of computation. Engineers who build bridges, physicists who compute temperatures over the surface of an unevenly heated plate, and accountants who work with the tax laws all do a lot of computational math. Some people like that.

Some parts of mathematics involve very little computation, such as graphing lines in the y = mx + b form (put a dot on the y-axis, draw the slope triangle, and slash in the line). The meaning of functions and the other concepts of this chapter (range, 1–1, onto, inverse function, identity function) require no calculator. The math fields of logic (If it rains, then the streets are wet. Therefore, if the streets are dry, it isn't raining.) or of set theory ({2, 7, 8} ∪ {7, 9} = {2, 7, 8, 9}) do not require that you know the multiplication tables. Some people like that. These non-computational fields don't require computing. However, they require *understanding*.

Chapter 7

Range of a Function

The range of a function is the set of images in the codomain. It is the set of all elements in the codomain that have a pre-image in the domain. The range is the set of all members of the codomain that are "hit" by at least one element of the domain.

510. What is the range of f?

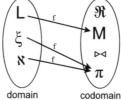

619. If the domain and codomain of g are the integers { ... –3, –2, –1, 0, 1, 2, 3, 4 ...} and g is defined by g(x) = |x|, what is the range of g?
The absolute value of x is defined in the fancy algebra books as $|x| = \sqrt{x^2}$. This does not make sense to a lot of people. Instead, it is easier to think of the absolute value of x as, "If x is a negative number, strip away its minus sign." Thus $|-8| = 8$, $|398| = 398$, and $|0| = 0$.

650. Can you invent a function f with a domain equal to the natural numbers {1, 2, 3, 4, 5 ...} whose range is {2, 4, 6, 8, 10, 12 ...}?

745. Invent a function g with domain equal to the whole numbers {0, 1, 2, 3, 4, 5 ...} whose range is the integers { ... –3, –2, –1, 0, 1, 2, 3, 4 ...}.
    This is a harder problem.
    g(x) = x won't work. None of the negative integers will be in the range.
    g(x) = –x won't work. Then numbers such as 5 or 3923 won't be in the range.
    Start by assigning 0 to 0. Then assign 1. Then assign 2. Then assign 3. Do it in such a way that none of the integers get left out. Please spend at least five minutes before you look at my answer.

Functions Represented as Ordered Pairs

Instead of writing $2 \xrightarrow{f} 5$, $3 \xrightarrow{f} 7$, $4 \xrightarrow{f} 19$, or writing f(2) =5, f(3) = 7, f(4) = 19, there is another way to represent this function, namely, (2, 5), (3, 7), (4, 19). The domain is {2, 3, 4} and the range is {5, 7, 19}.

66. Which of these are functions?
    Example A: (8, c), (32, d), (40, d)
    Example B: (2, r), (5, 5), (5, 3)
    Example C: (1, 2), (2, 3), (3, 4), (4, 5), (5, 6)
    Example D: (z, 123), (y, 50), (50, y)
    Example E: (v, m), (v, n), (v, p)

Chapter 7

114. For a moment, consider ordered pairs of *numbers*. If you graph two ordered pairs that have the same first coordinate (the same abscissa), what does the graph look like?

349. Which of these are graphs of functions?

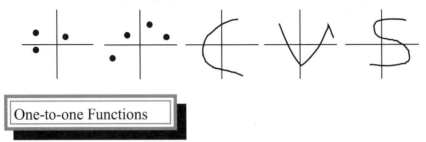

One-to-one Functions

Review: You are given two sets. Call the first set the domain and the second set, the codomain.
A function is any rule that assigns each element of the domain to exactly one element in the codomain.
New stuff: A function is 1-1 if no two elements in the domain are assigned to the same element in the codomain.

446. Let the domain and codomain of function f be the integers $\{\ldots -3, -2, -1, 0, 1, 2, 3, 4 \ldots\}$. Define f by $f(x) = x^2$.
 Is f a 1-1 function?

515. Addition is a function. The domain is the set of all ordered pairs of numbers and the codomain is the set of numbers. If we call the name of this function "+", then we can write:

$$(3, 5) \xrightarrow{+} 8$$
$$(-4, 7) \xrightarrow{+} 3$$
$$(800, 100) \xrightarrow{+} 900 \text{ etc.}$$

Is addition a 1-1 function?

584. Division is usually considered a function. However, if the domain were the set of all ordered pairs of kids, then division would not be a function. For something to be a function, each element in the domain must have exactly one image (one answer) in the codomain. But if (Sandy, Shelly) were in the domain, Sandy ÷ Shelly would have no meaning.

 Explain why division with a domain of the set of all ordered pairs of numbers would *not* be a function.

665. Describe the largest possible domain for the division function.

Chapter 7

Functions That Are Onto

Review: A function is 1-1 if no two elements in the domain have the same image.
New stuff: A function is onto its codomain if every element of the codomain is the image of at least one element of the domain.

98. Explain whether f is 1-1.
   Explain whether f is onto.

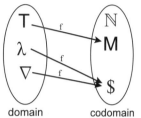

1240. The set of all elements in the codomain that are the image of at least one element in the domain is called the _____ of the function.
(fill in one word)

1366. The cardinality of a set is the number of elements in the set. For example, the cardinality of $\{ \Sigma, \%, 8, \circledast \}$ is 4.
   Suppose g is a function from set A to set B. Suppose g is onto B. Suppose the cardinality of set A is 28. What can you say about the cardinality of set B?

Inverse Functions

f:A → B means three things: ① f is the name of the function, ② the domain of f is A, and ③ the codomain of f is B.

74. Let g:A → B be the function defined by $g(x) = x^2$ and A = {3, 5, 9} and B = {9, 25, 81}.
   Then g(3) = 9, g(5) = 25, and g(9) = 81. Everybody is happy. ☺

   We now define a new function called "g inverse" (g^{-1}) which allows the codomain to fire back at the domain.
   We write g^{-1}:B → A. It's no surprise that $g^{-1}(9) = 3$. This is true because g(3) = 9.
   Your question: $g^{-1}(25) = $?
   Your second question: $g(g^{-1}(81)) = $? This means that we start with 81. Then we take g inverse of 81. Then we take g of that.

85. If $h(\pi) = \sqrt{2}$, then what does $h^{-1}(\sqrt{2})$ equal?

116. If f is defined by f(x) = x + 7, what does $f^{-1}(12)$ equal?

Chapter 7

192. If h is defined as $h(x) = (x + 4)^3$, what does $h^{-1}(8)$ equal?

365. Let $f: N \to N$ be defined by $f(x) = x + 7$ where N is the set of natural numbers $\{1, 2, 3, 4, 5, 6 \ldots\}$. Show that $f^{-1}: N \to N$ is *not* a function.

small essay

One-to-One, Onto, and Inverse

There are different types of mathematicians, just as there are different types of artists.

Some mathematicians like to do problems that require a lot of computation. They enjoy line after line of equations. They like figuring out how much steel is needed to make —

Those individuals often make excellent engineers.

There are other mathematicians (including me) who are called pure mathematicians in contrast to applied mathematicians. All of us—pure and applied—take the first two years of college math (called calculus) together. After that, as juniors in college, our paths separate.

Look into a classroom of future engineers and the blackboard will often have lots of calculations on it.

In contrast, proving things are true is a large part of pure mathematics. The only high school course that is chiefly pure math is geometry. The heart of geometry is proofs.

When I was a senior in college, in one pure math class our homework assignment one night was a single problem:

> If f: A → B is one-to-one
> and g: B → A is one-to-one,
> show that there exists a
> function h: A → B that is both one-to-one and onto.

Maybe two students were able to do it. (I wasn't one of them.)

end of small essay

Chapter 7

Second part: the 𝔐ixed 𝔅ag: a variety of problems from this chapter and previous material

56. Simplify $\log_{0.5} 16$ $\log 0.1$ $\log 1$

136. What is the distance between (7, 8) and (−2, 9)?

93. Graph $\dfrac{x}{2} + \dfrac{y}{2\pi} = 1$

221. Solve $\begin{cases} 4x - 3y = 10 \\ 7x + 2y = 3 \end{cases}$

231. Place into standard form for an ellipse
$49x^2 - 196x + y^2 + 6y + 156 = 0$

359. Graph $\dfrac{(x-3)^2}{49} - \dfrac{(y-4)^2}{1} = 1$

473. Graph $y > 3x^2 + 2x$

580. Let the domain and the codomain be the **real numbers**. (The real numbers are every number that is either positive, negative, or zero. They are every number on the number line.)

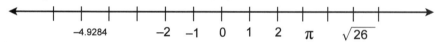

The real numbers do not include the imaginary numbers such as 5i.

Define f by the rule $f(x) = -x$.
Is f a function? Is f 1-1? Is f onto?

669. Suppose h is a function that maps set C (the domain) onto set D (the codomain). Suppose that the cardinality of set D is 6. What can you say about the cardinality of set C?

690. If f is a function defined by $f(x) = \log_7(x)$, what is $f^{-1}(x)$? This is not a super-easy question.

1380. Factor $2y^2 + 26y + 84$
$x^4 - 81$
$20x^2 + 11x - 3$

Chapter Seven and a Half

Problems on Each Topic

The Degree of a Polynomial

The degree of a term is the sum of the degrees of the variables in the term.

589. What is the degree of each of these terms?

$$67x^2y^7$$
$$3.082xyz$$
$$5y$$
$$\pi x$$

The degree of a polynomial is the degree of the highest-degree term in that polynomial.

670. What is the degree of each of these polynomials?

$$6x^6 + 43298x - 27xyz$$
$$\pi^2 x^4 + x^5$$
$$50000x - 0.0003x^3 + \sqrt{7}\,x^2$$

Long Division of Polynomials

Before dividing, arrange the terms of each polynomial in descending degree.

141. $29x^3 + 4x^2 + 10x^4 - 7x + 20 \div 2x + 5$

After arranging the terms in order of decreasing degree, if there are any terms left out, insert them with a zero coefficient. For example, turn $3x^5 - 7x^2$ into $3x^5 + 0x^4 + 0x^3 - 7x^2 + 0x + 0$.

521. $6x^5 + 7x - 12x^4 - 14 \div x - 2$

The only remaining detail is what to do when the remainder is not zero. In arithmetic The remainder was expressed as a fraction.

$$5\overline{)17} \quad 3\tfrac{2}{5}$$
$$\underline{15}$$
$$2$$

779. $\dfrac{25x^4 + 12x^5 + 7x^3 + 8 + 15x^2 + 14x}{3x + 1}$

To simplify this fraction by ① factor top, ② factor bottom, and ③ cancel like factors would be very hard. Instead, using long division makes the solution long, but not hard.

      Example of long, but not hard: Eating 40 gallons of ice cream. It may take weeks.

      Example of hard, but not long: Telling the truth when you are tempted to lie.

Chapter Eight

First part: Problems on Each Topic

Partial Fractions

Partial fractions is a way to "unadd" fractions. You start with the sum and find out what adds up to that sum. There are four possibilities:

Case 1: The denominator consists of distinct linear factors $\dfrac{8x - 29}{(x + 2)(x - 7)}$

Case 2: The denominator consists of linear factors with some of them repeated $\dfrac{5x^2 - 2x - 69}{(x - 7)(x + 2)^2}$

Case 3: The denominator consists of distinct quadratic factors $\dfrac{15x^3 - 5x^2 - 76x - 20}{(x^2 + 7x + 2)(3x^2 - 4)}$

Case 4: The denominator consists of quadratic factors with some of them repeated $\dfrac{12x^5 + 20x^4 + 120x^3 + 111x^2 + 251x + 25}{(x^2 + 5)^2(4x^2 + 3x)}$

The Chart

the first step for each of the cases
(A, B, C, D, E, F are numbers.)

Case 1: Distinct linear factors $\dfrac{8x - 29}{(x + 2)(x - 7)} = \dfrac{A}{x + 2} + \dfrac{B}{x - 7}$

Case 2: Repeated linear factors $\dfrac{5x^2 - 2x - 69}{(x - 7)(x + 2)^2} = \dfrac{A}{x - 7} + \dfrac{B}{x + 2} + \dfrac{C}{(x + 2)^2}$

Case 3: Distinct quadratic factors

$$\dfrac{15x^3 - 5x^2 - 76x - 20}{(x^2 + 7x + 2)(3x^2 - 4)} = \dfrac{Ax + B}{x^2 + 7x + 2} + \dfrac{Cx + D}{3x^2 - 4}$$

Case 4: Repeated quadratic factors

$$\dfrac{12x^5 + 20x^4 + 120x^3 + 111x^2 + 251x + 25}{(x^2 + 5)^2(4x^2 + 3x)} = \dfrac{Ax + B}{x^2 + 5} + \dfrac{Cx + D}{(x^2 + 5)^2} + \dfrac{Ex + F}{4x^2 + 3x}$$

Chapter 8

318. Unadd $\dfrac{8x - 29}{(x + 2)(x - 7)}$ Fancy algebra books might say, "Resolve into partial fractions."

390. Resolve into partial fractions (I can be fancy too!) $\dfrac{-2x - 26}{x^2 - 4x - 21}$

450. Unadd $\dfrac{46x - 3}{15x^2 - x - 2}$

564. Resolve into partial fractions $\dfrac{5x^2 - 2x - 69}{(x - 7)(x + 2)^2}$

839. Unadd $\dfrac{15x^3 - 5x^2 - 76x - 20}{(x^2 + 7x + 2)(3x^2 - 4)}$

975. Just do the first step (as the chart on the previous page shows) for each of these partial fractions problems. Do not go further.

$$\dfrac{5x^4 + 2x + 88}{(2x + 7)^2(x - 3)}$$

$$\dfrac{6x^3 + 5x + 34}{(9x - 1)(x^2 + 3)^2}$$

$$\dfrac{7}{x(2x^2 + 3)^3}$$

Linear Programming

There are two steps: First graph the inequalities on a single graph. Second, test each vertex.

53. Find the largest value that $f(x, y) = 3x + 4y$ becomes subject to the constraints: $x \geq 0$, $y \geq 0$, $3x + 5y \leq 45$, $6x + 5y \leq 60$

327. I lost my hose so I use water from the tub and from the kitchen sink to water my lawn.

    Tub water contains 3 parts of iron and 1 part of sulfur per gallon.

    Sink water contains 2 parts of iron and 4 parts of sulfur per gallon.

    My lawn needs at least 18 parts of iron and 8 parts of sulfur. (You can see from the picture that my lawn is not a happy lawn.)

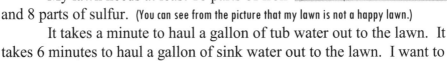

    It takes a minute to haul a gallon of tub water out to the lawn. It takes 6 minutes to haul a gallon of sink water out to the lawn. I want to minimize the amount of time used in hauling water.

Hints & Help: We have had very few word problems in advanced algebra. English is harder than math. Getting to the inequalities is a lot harder than doing the graphing and testing the vertices. The first step in

Chapter 8

any word problem is to "Let x equal" and "Let y equal." You do that by reading the English and finding out what we are looking for.

You know we are trying to minimize the time spent, but what are we looking for? We need to know how many gallons of tub water and how many gallons of sink water we will use.

Let x = number of gallons of tub water used.

My garage when I first moved in

397. When I first moved into my house years ago, I knew that I had to store some of my math books and shoes in the garage. The red moving box could hold 12 math books and 6 shoes. The blue moving box could hold 12 math books and 10 shoes.

I knew that I needed to store at least 96 math books and 60 shoes. Should I use red boxes, blue boxes, or some of each? Red boxes take up 9 cubic feet. Blue boxes, 10. I want to minimize the volume.

600. The railing was a mess when I bought the house. Each gallon of Flubber paint takes 2 hours to apply and will cover 100 square feet. Each gallon of Gauss paint takes 1 hour to apply and will cover 200 square feet.

I have, at most, 10 hours to do the painting and I need to cover 1,000 square feet of railing.

Flubber paint costs $12/gallon and Gauss paint costs $90/gallon. How much of each should I use in order to minimize the cost?

895. There were a lot of rocks in my backyard that I needed to remove. I had two options. I could either use a hammer and wheelbarrow or use dynamite and a truck. (This problem was mentioned in the introduction of this book.)

With the hammer and wheelbarrow, my cost would be $1/day and I would get injured about 0.2 times each day. (On the average, I would get injured once every 5 days.)

With the dynamite and a truck, my cost would be $50/day and I would get injured about 0.1 times each day.

I had $817 to spend on the project and I could experience at most 5 injuries. (After five injuries I would quit and do something a lot less dangerous. Being a mathematician is one of the safest jobs there is.)

I can remove 4 tons of rocks each day using a hammer and wheelbarrow and 20 tons using dynamite and a truck. What is my best course of action in order to remove as much rock as possible?

Chapter 8

1090. Here is a row of history books and a row of poetry books from my library.

Each page of my history books has 3 facts, stirs 2 good emotions in me, and takes 4 calories of effort to read.

Each page of my poetry books gives me 1 fact, stirs 7 good emotions in me, and takes 2 calories of effort to read.

Tonight I want to get at least 12 facts from my reading and at least 65 good emotions. I want to spend the minimum number of calories to do this. How should I divide my time between the history and poetry books?

1167. Repeating some of the given from the previous problem: Each page of my history books has 3 facts, stirs 2 good emotions in me, and takes 4 calories of effort to read. Each page of my poetry books gives me 1 fact, stirs 7 good emotions in me, and takes 2 calories of effort to read.

I have just eaten a tiny piece of chocolate and have 20 calories to spend on reading. I still want to get at least 12 facts from my reading.

How should I divide my time between the history and poetry books in order to maximize my good emotions?

Math Induction Proofs

Math induction allows you to prove that an infinite number of different statements are true.

For example, $2 + 4 + 6 + 8 + \ldots + 2n = n(n + 1)$ is true for ...
$\quad\quad n = 1 \Rightarrow 2 = 1(1 + 1)$
$\quad\quad n = 2 \Rightarrow 2 + 4 = 2(2 + 1)$
$\quad\quad n = 3 \Rightarrow 2 + 4 + 6 = 3(3 + 1)$
$\quad\quad n = 4 \Rightarrow 2 + 4 + 6 + 8 = 4(4 + 1)$
$\quad\quad$ etc.

To prove all of these statements true (by math induction) you do two things:
$\quad\quad$ First, you prove that the $n = 1$ statement is true.
$\quad\quad$ Second, you assume that the $n = k$ statement is true and you prove that the $n = k+1$ statement must be true.

329. Prove $1 + 3 + 5 + 7 + \ldots + 2n-1 = n^2$ for every natural number n. (The natural numbers = $\{1, 2, 3, 4 \ldots\}$)

Chapter 8

370. Prove $1^3 + 2^3 + 3^3 + 4^3 + \ldots + n^3 = \dfrac{n^2(n+1)^2}{4}$ is true for every natural number.

small essay

Induction

The proof by induction taught in *Life of Fred: Advanced Algebra Expanded Edition* is the only type of induction proof that is mentioned in high school math.

You have an infinite number of statements: $S_1, S_2, S_3, S_4 \ldots$ that you want to show are true. You prove S_1 is true. You assume that S_k is true and then show that S_{k+1} must be true.

But that is not the end of the story of induction.

In later math courses, we have a second form of proof by induction. It is called **strong induction**. You have an infinite number of statements: $S_1, S_2, S_3, S_4 \ldots$ that you want to show are true. You prove S_1 is true. You assume that $S_1, S_2, S_3, S_4, \ldots, S_{k-1}, S_k$ are all true and then show that S_{k+1} must be true.

In later math courses, we have a third form of proof by induction. It is called **transfinite induction**. Why stop with $S_1, S_2, S_3, S_4 \ldots$? In Chapter 1½, in the answer to problem #1077, we mentioned numbers that are larger than any natural number (= $\{1, 2, 3, 4, 5, \ldots\}$) and had names like $\aleph_1$ and $\aleph_2$.

With transfinite induction we will prove more than just a simple infinite list of statements is true.

end of small essay

Chapter 8

Second part: the 𝔐ixed 𝔅ag: a variety of problems from this chapter and previous material

245. How long (L) I have to wait in line at the amusement park varies directly as the square root of the number of kids (n) at the park.
 If there are 20 kids, I have to wait in line 7 minutes. How long is the wait if there are 180 kids?

275. Solve $\sqrt{y^2 - 24} = 5$

291. What is the equation of the line that passes through (32, –4) and (30, 9)?

323. Graph $y = \frac{3}{4}x - 2$

533. $\dfrac{5x^2 + 18x + 9}{9x^2 + 6x - 8} \div \dfrac{x^2 - 9}{3x^2 - 11x + 6}$

211. Solve $\dfrac{x}{x+2} = \dfrac{6}{x^2 + x - 2}$ Hint: Solve as a fractional equation rather than cross-multiplying. If you cross-multiply, you will get a cubic equation (an equation containing x^3)

236. Find the equation of the circle the circumscribes the ellipse
$$\frac{(x-3)^2}{36} + \frac{(y-5)^2}{64} = 1$$
The circle that circumscribes an ellipse is the smallest possible circle that contains the ellipse.

419. Solve $\begin{cases} 4x + 3y = 8 \\ 6x + y = 5 \end{cases}$

445. Is this a function? Is it 1-1? Is it onto?

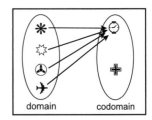

691. Resolve into partial fractions
$$\frac{6x^2 + 17x - 13}{(x+1)(x+3)(x-2)}$$

1492. If g is a function defined by $g(x) = 5^x$, what is $g^{-1}(x)$ equal to?

Chapter 8

517. Describe the largest possible domain for the function whose rule is $f(x) = \sqrt{x}$.

1180. On a day I go barefoot, my arches improve by 3 points, I get 8 ouches, and make 5 new friends.

On a day I wear shoes, my arches improve by 1 point, I get 2 ouches, and make 4 new friends.

I want my arches to improve by at least 18 points and I want to get no more than 24 ouches. How many days should I go barefoot and how many days should I wear shoes in order to make as many new friends as possible?

1228. Repeating some of the given from the previous problem: On a day I go barefoot, my arches improve by 3 points, I get 8 ouches, and make 5 new friends. On a day I wear shoes, my arches improve by 1 point, I get 2 ouches, and make 4 new friends.

I want my arches to improve by at least 18 points and I want to make at least 40 new friends.

How many days should I go barefoot and how many days should I wear shoes in order to minimize my ouches?

1313. Graph $y = 3x^2 - 5$ from $x = -2$ to $x = 2$.

1422. Prove $9 + 13 + 17 + 21 + \ldots + 4n + 5 = n(2n + 7)$ is true for every natural number.

Chapter Nine

First part: Problems on Each Topic

Arithmetic Sequences and Series

4, 11, 18, 25, 32, 39, 46, 53 is an arithmetic sequence.
4 + 11 + 18 + 25 + 32 + 39 + 46 + 53 is an arithmetic series.
The first term, a, in this case is 4.
The last term, ℓ, in this case is 53.   I am going to use a *cursive* ℓ so it won't be confused with 1 (one).
The common difference between terms, d, in this case is 7.
The number of terms, n, in this case is 8.
The sum of the series is s.

The two formulas are: $\ell = a + (n-1)d$ and $s = \frac{n}{2}(a + \ell)$

36. What is the 77th term of $5 + 9 + 13 + 17 + 21 + \ldots$?

814. How many terms are in the sequence 23, 26, 29, 32, . . . , 1949 ?

196. On the first day of the month, I spent $15 on bear food. On the second day, I spent $25. On the third day, $35. Each day I spent $10 more than the previous day. (The bears got hungrier and hungrier.)
 How much did I spend in the first 30 days?

Geometric Sequences and Series

The plural of *sequence* is *sequences*. The plural of *series* is *series*.
I don't think *English* has a plural.
When you are talking about a word, you put it in italics.   Iowa is a state. *Iowa* has four letters.
4, 12, 36, 108, 324 is a geometric sequence.
4 + 12 + 36 + 108 + 324 is a geometric series.

a = 4 is the first term. ℓ = 324 is the last term. s is the sum of the geometric series.
r = 3 is the common ratio between the terms. To get from 4 to 12 you multiplied by 3. To get from 12 to 36 you multiplied by 3. To get from 36 to 108 you multiplied by 3.

$$\ell = ar^{n-1} \quad \text{and} \quad s = \frac{a(1-r^n)}{1-r}$$

Chapter 9

301.

My dog Wufwuf has 4 fleas on the day I have given her a bath. We'll call that day #1. On the next day (day #2) she has 28 fleas and the day after that (day #3), 196 fleas. The sequence is 4, 4·7, 4·7², 4·7³, ...
(4·7 means 4 times 7)

How many fleas does Wufwuf have on day #22?

330. Most dogs like to fetch balls. Wufwuf is different. She likes to fetch pumpkins. One Halloween she ran around the neighborhood and stole 3 pumpkins and brought them home. She didn't tell me about her theft but hid the pumpkins behind the backyard wall.

On the second Halloween, she stole 12 pumpkins and added them to her stash behind the backyard wall.

On the third Halloween, she stole 48 pumpkins. Each year she got better at finding pumpkins and taking them. Each year she took 4 times as many as the previous year.

How many pumpkins were hidden behind the backyard wall after 8 Halloweens? (In case you are wondering, Wufwuf only stole plastic pumpkins. They didn't rot.)

Sigma Notation   Σ

$$\sum_{i=1}^{5} 8ix = 8x + 8(2)x + 8(3)x + 8(4)x + 8(5)x$$

$$\sum_{i=1}^{3} 7x^i = 7x + 7x^2 + 7x^3$$

77. Write out what each of these means:

$$\sum_{i=1}^{8} 5i$$

$$\sum_{i=1}^{4} \log(x + i)$$

$$\sum_{i=1}^{3} 8x^i y^{i+3}$$

Chapter 9

749. Which of these are arithmetic series and which are geometric?

$$\sum_{i=1}^{6} (9+i)$$

$$\sum_{i=1}^{3} 7i$$

$$\sum_{i=1}^{9} 4^i$$

1429. Give an example of a series that is both arithmetic and geometric.
There are an infinite number of possible examples.

350. Express in sigma notation $\log(3x^2) + \log(3x^3) + \log(3x^4) + \log(3x^5)$

Every infinite arithmetic series will have an infinite sum.
    $8 + 8 + 8 + 8 + \ldots$ will get larger than any number you can name.
The only exception is $0 + 0 + 0 + \ldots$.

Every infinite geometric series (where $r > 1$) will have an infinite sum.
    $3 + 3(1.07) + 3(1.07)^2 + 3(1.07)^3 + \ldots$ will get larger than any number than you can name.
Or if $r < -1$ we get a series whose sum is unbounded. $5 - 15 + 45 - 135 + \ldots$ $a = 5$ and $r = -3$

The only happy case is a geometric series in which r is between -1 and 1. If $-1 < r < 1$, then the sum of the infinite number of terms is finite.

$$s = \frac{a}{1-r}$$ For many, this is their favorite formula in algebra.

---

963. Find the sum of $\sum_{i=1}^{\infty} 2(0.4^i)$

1022. My favorite pizza place is always open. It is 6 a.m. and I need my breakfast pizza. (Is there a better way to start the day?) I run from my house toward the pizzeria and cover half of the distance in 8 minutes. I continue running and cover half of the remaining distance in 4 minutes. I cover half the remaining distance in 2 minutes and so on. What time will it be when I get there?

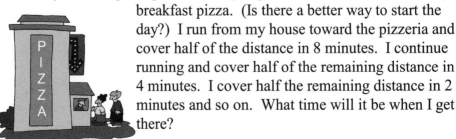

Chapter 9

Matrices

879. Adding matrices is sleepy-time easy.

Suppose that the inventory in my Alabama store is given by the matrix

$$A = \begin{pmatrix} 3 & 5 & 11 & 2 \\ 4 & 7 & 8 & 6 \\ 1 & 0 & 23 & 7 \end{pmatrix} \begin{matrix} \leftarrow \text{pairs of children's shoes} \\ \leftarrow \text{pairs of women's shoes} \\ \leftarrow \text{pairs of men's shoes} \end{matrix}$$

pink　red　black　brown

In the language of matrices, this matrix is a 3 × 4 ("three by four") matrix. It has three rows and four columns.

The entry in the second row, third column is 8. That indicates that my Alabama store has 8 pairs of black women's shoes in its inventory.

The inventory in my Colorado store is given by

$$C = \begin{pmatrix} 5 & 1 & 22 & 3 \\ 0 & 2 & 14 & 5 \\ 0 & 0 & 38 & 4 \end{pmatrix} \begin{matrix} \leftarrow \text{pairs of children's shoes} \\ \leftarrow \text{pairs of women's shoes} \\ \leftarrow \text{pairs of men's shoes} \end{matrix}$$

pink　red　black　brown

The combined inventory of the two stores is found by adding the matrices.

Compute A + C.

1095. Multiplying matrices is not sleepy time easy nor is it Einstein hard. It's just a little awkward. As one student once explained to me: Your left eye wants to go left-right ↔ and your right eye wants to go up-down ↕.

Here is the 1×4 matrix of the children's shoes in my Alabama store: (3　5　11　2) where the columns are pink　red　black　brown. Suppose the price of a pair of child's pink shoes is $7, a pair of child's red shoes is $6, a pair of child's black shoes is $4, and a brown child's shoe is $5.

The three pairs of pink shoes cost a total of $21. The five pairs of red shoes, $30.
The inventory matrix is (3　5　11　2).

The price matrix is a 1×4 matrix $\begin{pmatrix} 7 \\ 6 \\ 4 \\ 5 \end{pmatrix}$

Chapter 9

We multiply the inventory matrix times the price matrix:

$$(3 \quad 5 \quad 11 \quad 2) \begin{pmatrix} 7 \\ 6 \\ 4 \\ 5 \end{pmatrix} = (105)$$

The 105 is $(3)(7) + (5)(6) + (11)(4) + (2)(5)$

Did you notice that your left eye was moving ↔ and your right eye was moving ↕?

The $105 represents the cost of all pairs of the children's shoes in the Alabama store.

The question for you: Suppose there is a price increase and the price of all pairs of pink, red, black, and brown shoes are, respectively, $9, $11, $8, $10. Find the cost of all the women's shoes in my Alabama store.

In other words, multiply $(4 \quad 7 \quad 8 \quad 6) \begin{pmatrix} 9 \\ 11 \\ 8 \\ 10 \end{pmatrix}$

1264. Now compute the total cost of the children's, women's and men's shoes in my Alabama store.

$$\begin{pmatrix} 3 & 5 & 11 & 2 \\ 4 & 7 & 8 & 6 \\ 1 & 0 & 23 & 7 \end{pmatrix} \begin{pmatrix} 9 \\ 11 \\ 8 \\ 10 \end{pmatrix} = (\quad ? \quad 237 \quad ? \quad)$$

pink red black brown

To get the first ?, you multiply $(3 \quad 5 \quad 11 \quad 2)$ times $\begin{pmatrix} 9 \\ 11 \\ 8 \\ 10 \end{pmatrix}$

To get the second ?, you multiply $(1 \quad 0 \quad 23 \quad 7)$ times $\begin{pmatrix} 9 \\ 11 \\ 8 \\ 10 \end{pmatrix}$

Chapter 9

1318. $\begin{pmatrix} 6 & 0 & -2 \\ 2 & 1 & 16 \end{pmatrix} \begin{pmatrix} 3 & 5 & 2 \\ 4 & 1 & 0 \\ 2 & 0 & 1 \end{pmatrix}$

1075. Which of these pairs of matrices can be multiplied together as they are written?

Example A: $\begin{pmatrix} 8 & 4 & 4 & 3 \\ 3 & 8 & 7 & 5 \\ 9 & 0 & 1 & 2 \end{pmatrix} \begin{pmatrix} 8 & 4 & 4 & 3 \\ 3 & 8 & 7 & 5 \\ 9 & 0 & 1 & 2 \end{pmatrix}$

Example B: $\begin{pmatrix} 5 & 33 & 763 \\ 3 & -8 & 23 \end{pmatrix} \begin{pmatrix} 3983 \\ 81 \\ 3974 \end{pmatrix}$

Example C: $(2 \quad 5 \quad 0 \quad -3) \begin{pmatrix} 8311 \\ 9832 \\ 4320 \\ 6 \end{pmatrix}$

Example D: $\begin{pmatrix} 7 \\ 8 \\ 2 \end{pmatrix} (44398 \quad 98692 \quad 23249)$

Chapter 9

Second part: the 𝔐ixed 𝔅ag: a variety of problems from this chapter and previous material

353. Simplify $\log_{36} 6$
$\log_{\sqrt{7}} 7$
$\log_{1000} 10$

649.   Three geese and one boy eat 1.1 pounds of food each day.
    Five geese and three boys eat 2.5 pounds of food each day.
    Let x = lbs. of food eaten by one goose.
    Let y = lbs. of food eaten by one boy.
    How much food does one goose eat each day?
    How much food does one boy eat each day?

774. Graph $-\dfrac{(x+7)^2}{16} + \dfrac{(y-5)^2}{4} > 1$

835. The domain is the set of whole numbers {0, 1, 2, 3, 4, . . .}, and the codomain is the set of natural numbers {1, 2, 3, 4, . . .}. Define f by the rule f(x) = 2x. Is f a function?

995. Let g be a function whose domain is {a, b, c, d} and whose codomain is {a, b, c, d, e, f }. Could g be 1-1? Could g be onto?

1130. Find the sum of each of these:

$$\sum_{i=1}^{44} 6+i$$

$$\sum_{i=3}^{72} 4i$$

$$\sum_{i=1}^{\infty} \left(\frac{3}{8}\right)^i$$

1217. $\begin{pmatrix} 3 & 9 \\ 5 & 2 \\ -2 & 4 \end{pmatrix} \begin{pmatrix} 3 & 2 & 4 & 9 \\ 9 & 1 & 1 & 7 \end{pmatrix}$

1234.   Find the mean and median averages of 300, 700, 500.

Chapter 9

1025.

While vacationing in sunny Sandeneyes, I visited their famous Pottery Shop.
  On the top row were round pots that cost $3 each, weighed 5 pounds, and could hold 2 liters of root beer.
  On the bottom row were tall, skinny pots that cost $6 each, weighed 15 pounds, and could hold 5 liters of root beer.
  My budget was $30 for buying pottery. I could carry at most 60 pounds. How many of each should I buy so that I could fill them with as much root beer as possible?

1315. Let the domain of function g be all real numbers. Define g by the rule $g(x) = \sqrt{x^2 + 2}$. Show that g is *not* 1-1.

1499. I accidentally left the water running in my backyard. In the first minute, it washed away 5 cubic inches of dirt. In the second minute, it washed away $5(0.92)$ cubic inches. In the third minute, $5(0.92)^2$ cubic inches. Each minute it washed away 92% as much as the previous minute.
  After 20 minutes, how many cubic inches of dirt had been washed away?

Chapter Ten

First part: Problems on Each Topic

The Fundamental Principle

If there are m ways to do one thing and n ways to do another, there are mn ways to do both.

59. There are four places I need to get to today: the grocery store, the bank, the office supply store, and the library. I could do them in this order: GS—B—OSS—L or in this order B—L—OSS—GS.
   How many different orders could I do them in?

67. (continuing the previous problem) Suppose I need to get to the grocery store last because I'm buying ice cream and it will melt if I leave it in the car too long. In how many different orders can I visit those four spots?

99. (continuing the previous problem) Suppose I decide to visit the four spots in this order: B—L—OSS—GS.
   At the bank I have to decide whether to withdraw $20 or $40 (2 choices). At the library I have to decide which section of the library to go to first. There are 10 possibilities: the math section, the poetry section, the religion section, the history section, the business section, etc. At the office supply store I have to decide whether I want to get another fountain pen (2 choices—yes or no). At the grocery store I will choose one of my four favorite ice cream flavors.
   Here is one possible story: I withdrew $20. I headed to the history section first, I looked at all the fountain pens but decided not to get one today. I chose chocolate ice cream.
   How many possible stories are there?

815. In Chapter 7 we introduced the notation f:A → B. This meant a function whose name is f, whose domain is set A, and whose codomain is set B.
   How many possible functions are there when f:$\{1, 2, 3\}$ → $\{7, 8\}$?

750. How many possible functions are there for g when we are given g:$\{1, 2, 3, 4, 5, 6, \ldots, 99, 100\}$ → $\{a, b, c\}$?

1408. How many possible functions are there for h when we are given h:A → B where A has 47 elements and B has 80 elements?

Chapter 10

977. How many possible functions are there for f if f:A → B where A has 6 members and B has 8 members and f is 1-1? (A function is 1-1 if no two members of the domain have the same image.)

1455. How many possible functions are there for g if g:A → B where A has 100 members and B has 300 members and g is 1-1?

$P(n, r)$ The Permutation of n Items Taken r at a Time

1187. There are 300 flavors of ice cream. A survey was taken to determine the 100 most popular flavors. The results of the survey revealed the first place winner, the second place winner, and so on, all the way down to the 100th most popular flavor.

How many ways could this survey have turned out?

1321. Ice cream! The menu suggested either a single scoop or a double scoop. How silly. They have nine flavors so I want nine scoops, one of each flavor.

It matters what order the flavors are arranged on the cone. How many different ways could they arrange those nine flavors?

86. Without using a calculator, find the value of $\dfrac{24!}{23!4!}$

$C(n, r)$ The Combination of n Things Taken r at a Time

Combinations = choosing r things out of n things where the order does not matter.

$$C(n, r) = \dfrac{n!}{(n-r)!\,r!}$$

47. Here is the recipe for Triple-Scoop Milkshake

1 pint whipping cream
1 cup peach sauce
3 scoops ice cream—all different flavors
Blend until smooth. Serve with a spoon and a cherry.

If you have 9 flavors of ice cream available, how many different Triple-Scoop Milkshakes are possible?

Chapter 10

60. My mother said I could have anything I wanted at the ice cream store. I had two of the nine-scoop ice cream cones and then had three of the Triple-Scoop milkshakes.

  I turned blue. My mother thought it was because I was feeling cold. I explained to her that it felt like an elephant was sitting on my chest. She told me I had a tummy ache.

  I said that it was my chest, not my stomach.

  The guy behind the counter said, "It's probably just a heart attack. We get them all the time here. The hospital is 4 blocks south and 5 blocks east of here."

  She threw me in the car and headed to the hospital. This is the route she took.

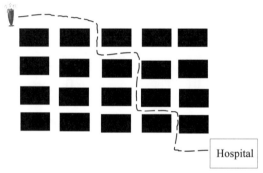

  This is a hard question. Only 1.3% of readers will be able to answer it. How many different routes are possible from the ice cream store to the hospital?

  Once you see my answer, you and 94% of all readers will say, "If you look at the question in the right way, it is pretty easy."

998. When my daughter, Jill, was learning her letters, she hopped on my lap when I was using a typewriter. She typed J   i   l   and then looked at me and asked, "Where is another l?"

  I explained to her that she could hit the l key a second time.

  How many 6-letter "words" are there? For example, jMEtmc and VVLqqq. Assume there are 52 possible letters (26 lowercase and 26 uppercase).

1394. Suppose you are playing Scrabble™ and you get these seven tiles: L L L P Q R S. How many different ways can you arrange them in a row?

Chapter 10

87. You are the director of the hospital. There are 12 doctors on the staff. You need to assign 5 of them to the emergency ward, 3 to pediatrics, and the remaining 4 will be assigned as lobbyists to Congress.

How many different ways could you make those assignments?

The Binomial Formula and Pascal's Triangle

$x + y$ is a binomial. So is $329x^6w + 928904yz^5$.

$(x + y)^2 = (x + y)(x + y) = x^2 + xy + xy + y^2 = x^2 + 2xy + y^2$

$(x + y)^3 = (x + y)(x^2 + 2xy + y^2) = x^3 + 2x^2y + xy^2 + x^2y + 2xy^2 + y^3 = x^3 + 3x^2y + 3xy^2 + y^3$

$(x + y)^4 = x^4 + 4x^3y + 6x^2y^2 + 4xy^3 + y^4$ (skipping steps)
$(x + y)^5 = x^5 + 5x^4y + 10x^3y^2 + 10x^2y^3 + 5xy^4 + y^5$

If we toss in $(x + y)^0 = 1$ and look at the coefficients for each of these, we have . . .

$(x + y)^0 =$ **1**
$(x + y)^1 =$ **1**$x +$ **1**y
$(x + y)^2 =$ **1**$x^2 +$ **2**$xy +$ **1**y^2
$(x + y)^3 =$ **1**$x^3 +$ **3**$x^2y +$ **3**$xy^2 +$ **1**y^3
$(x + y)^4 =$ **1**$x^4 +$ **4**$x^3y +$ **6**$x^2y^2 +$ **4**$xy^3 +$ **1**y^4
$(x + y)^5 =$ **1**$x^5 +$ **5**$x^4y +$ **10**$x^3y^2 +$ **10**$x^2y^3 +$ **5**$xy^4 +$ **1**y^5

These numbers (without the x's and y's) are the beginning of Pascal's Triangle. It goes on forever.
The next line would be **1 6 15 20 15 6 1**
The next line would be **1 7 21 35 35 21 7 1**

There is no need to memorize it. Each entry is the sum of the two entries above it. **21** is **6 + 15**

Now, the weird part. The numbers in Pascal's Triangle are C(n, r) entries.

Here is Pascal's Triangle:

$$\begin{array}{c}
C(0, 0) \\
C(1, 0) \quad C(1, 1) \\
C(2, 0) \quad C(2, 1) \quad C(2, 2) \\
C(3, 0) \quad C(3, 1) \quad C(3, 2) \quad C(3, 3) \\
\text{etc.}
\end{array}$$

73

Chapter 10

Hey! Now you can instantly write out what $(x + y)^{39}$ is equal to.

$(x + y)^{39} = C(39, 0)x^{39} + C(39, 1)x^{38}y + C(39, 2)x^{37}y^2 + C(39, 3)x^{36}y^3 +$ etc.

48. Using C's, write out the first five terms of $(x + y)^{57}$.

Let's make things easier.

Instead of writing
$(x + y)^{39} = C(39, 0)x^{39} + C(39, 1)x^{38}y + C(39, 2)x^{37}y^2 + C(39, 3)x^{36}y^3 + \ldots$

we could write
$(x + y)^{39} = x^{39} + \dfrac{39}{1!}x^{38}y + \dfrac{39 \cdot 38}{2!}x^{37}y^2 + \dfrac{39 \cdot 38 \cdot 37}{3!}x^{36}y^3 + \ldots$

If you really look at that previous line, you can see all kinds of patterns.

x^{39} is a pretty obvious way to start.
Then the "39" hops on top of the fraction bar.
The exponent on the x decreases by one. The exponent on the y increases by one.
The denominators are 1!, 2!, 3!, 4! . . . .
In the next term the "38", which was the exponent on the previous x, hops down and joins the 39.

The term after $\dfrac{39 \cdot 38 \cdot 37}{3!}x^{36}y^3$ is easy to predict: $\dfrac{39 \cdot 38 \cdot 37 \cdot 36}{4!}x^{35}y^4$

68. Write out the first five terms of $(x + y)^{74}$.

37. Write out the first five terms of $(7x^3 + 8yz^5)^{94}$.

Chapter 10

Second part: the 𝔐ixed 𝔅ag: a variety of problems from this chapter and previous material

45. If f is a function defined by $f(x) = x^3$ and its domain is {0, 2, 3}, what is the range of f ?

113. $(3 + 8i)(2 + 5i)$ i is the imaginary number that is equal to $\sqrt{-1}$

126. $\dfrac{\log_6 144}{\log_6 12} = ?$

156. What is the equation of the line with a slope of –4 that passes through the point (2, 9)?

191. Graph $y - \dfrac{2}{3}x > 5$

290. What is the 503rd term of 10 + 18 + 26 + 34 + 42 + . . . ?

300. The domain and the codomain are the natural numbers {1, 2, 3, 4, 5, 6, 7, . . .} and g is defined by the rule $g(x) = x^2$.
    Is g a function? Is g 1-1? Is g onto?

590. How many functions f are possible if f: {L, M, N} → {P, Q, R, S}?

654. Which of these are graphs of functions?

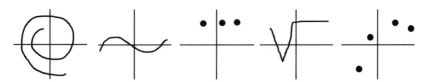

962. One way to describe a function is as a set of ordered pairs. (a, b) will mean the same thing as a $\xrightarrow{f}$ b.
    Define function h as {(3, 9), (5, 7), (–9, 8), (6, 7)}.
    What is the domain of h?
    What is the range of h?
    Is h 1-1?

974.   Example A: {(8, 8), (9, 2), (2, 9)} Is this a function?
    Example B: {(1, 2), (2, 4), (3, 6), (3, 8)} Is this a function?
    Example C: {(5, a), (b, a), (7, c), (9, d)} Is this function 1-1?
    Example D: {(M, L), (L, P), (P, Q), (Q, R)} Is this function 1-1?

1212. Resolve into partial fractions $\dfrac{7x^2 + 6x - 21}{(x-3)(x+2)(x+1)}$

Chapter 10

1258. On the first day of class, the music teacher told the students to practice their violins for a certain number of minutes that night. On the next day of class, he told them to practice for 4 minutes more. On the third day of class he told them to practice for 4 minutes more than the previous night. Each night they practiced 4 minutes longer than the previous night. On the 66th day of class, they were told to practice for 277 minutes that night. How long did they practice on the first night?

525.

We were so happy when the house mortgage was paid. It was a 30-year mortgage. (We made 360 monthly payments.)

The first payment paid $100 toward the principal balance. (The total amount owed on the principal went down by $100.)

The second payment decreased the principal balance by $107. The third payment decreased the principal balance by $114. Each monthly payment decreased the principal balance by $7 more than the previous month. By the end of the 360 payments, the balance was zero.

Translation: The 360 payments on the principal of 100, 107, 114, 121 . . . equaled the original mortgage. What was the original mortgage?

1297. Find the sum of each of these.

$$\sum_{i=1}^{8} (0.3)^i$$

$$\sum_{i=1}^{\infty} 5\left(\frac{1}{4}\right)^i$$

$$\sum_{i=5}^{55} (\pi + i)$$

1352. Graph $y = 3x^2 + 2x - 7$ from $x = -4$ to $x = 4$.

1387. They sang 8 Christmas carols and asked me to choose my favorite, my second favorite, and my third favorite. How many ways could I pick first, second, and third place favorites?

Chapter 10

1475. Planning a birthday cake is tough. Those star candles cost 2¢ each, give off 3 units of nice smell, and add 5 units of fun. (The fun is lighting the candles and blowing them out.)

The flowers cost 70¢ each, give off 8 units of nice smell, and add 2 units of fun.

We want at least 24 units of nice smell and at least 20 units of fun.

How can we do this and minimize the cost?

The Complete Solutions and Answers

35. Find the median average of 6, 8, 22, 49, 51.

To find the median average of a set of numbers, you first line the numbers up from smallest to largest. Then you pick the one in the middle. The numbers {6, 8, 22, 49, 51} are already sorted from smallest to largest. Just pick the one ↑ in the middle.

The median average is 22.

If you had to find the median height of a bunch of hamburgers, you could just line them up and pick the one in the middle.

36. What is the 77th term of $5 + 9 + 13 + 17 + 21 + \ldots$?

$$\begin{aligned} \text{The first term} \quad & a = 5 \\ \text{The common difference} \quad & d = 4 \\ \text{The number of terms} \quad & n = 77 \end{aligned}$$

$$\ell = a + (n-1)d = 5 + (76)4 = 5 + 304 = 309$$

37. Write out the first five terms of $(7x^3 + 8yz^5)^{94}$.

$(7x^3 + 8yz^5)^{94} =$

$$(7x^3)^{94}$$

$$+ \frac{94}{1!}(7x^3)^{93}(8yz^5) \quad \text{Once you have gotten this far, the rest is on automatic pilot.}$$

$$+ \frac{94 \cdot 93}{2!}(7x^3)^{92}(8yz^5)^2$$

$$+ \frac{94 \cdot 93 \cdot 92}{3!}(7x^3)^{91}(8yz^5)^3$$

$$+ \frac{94 \cdot 93 \cdot 92 \cdot 91}{4!}(7x^3)^{90}(8yz^5)^4$$

The last term will be $\frac{94!}{94!}(7x^3)^0(8yz^5)^{94}$ or more simply $(8yz^5)^{94}$

The Complete Solutions and Answers

38. What is the least and the greatest number of quadrants that a straight line might lie in?

Answer: No graph of a straight line can lie in all four quadrants.

Most straight lines lie in three quadrants.

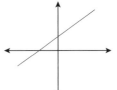

Some people might think that a straight line must lie in at least two quadrants,

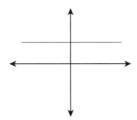

but they would be wrong. The x-axis, for example, also known as $y = 0$, is a straight line that does not lie in any of the quadrants.

41.

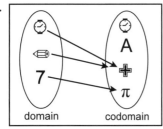

example 1

This is a function.
⊘ has one image in the codomain.
⊜ has one image in the codomain.
7 has one image in the codomain.

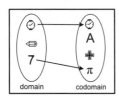

example 2

⊜ has zero images in the codomain.
This is not a function.

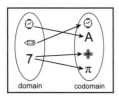

example 3

7 has two imagines in the codomain.
This is not a function.

| 44–48 | The Complete Solutions and Answers |

44. Find the slopes of lines ℓ, m, and n.

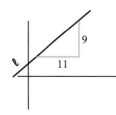 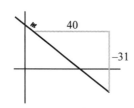 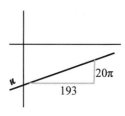

slope = $\dfrac{\text{rise}}{\text{run}} = \dfrac{9}{11}$ slope = $\dfrac{\text{rise}}{\text{run}} = \dfrac{-31}{40}$ slope = $\dfrac{\text{rise}}{\text{run}} = \dfrac{20\pi}{193}$

45. If f is a function defined by $f(x) = x^3$ and its domain is $\{0, 2, 3\}$, what is the range of f ?

$f(0) = 0$ $f(2) = 8$ $f(3) = 27$ so the range of f is $\{0, 8, 27\}$

47. Here is the recipe for Triple-Scoop Milkshake

1 pint whipping cream
1 cup peach sauce
3 scoops ice cream—all different flavors
Blend until smooth. Serve with a spoon and a cherry.

If you have 9 flavors of ice cream available, how many different Triple-Scoop Milkshakes are possible?

The order that the three flavors of ice cream are chosen does not matter. They will all be blended together.

$$C(n, r) = \dfrac{n!}{(n-r)!r!} \quad \text{becomes} \quad C(9, 3) = \dfrac{9!}{6!3!}$$

Since 9!/6! = 9×8×7 $\dfrac{9!}{6!3!} = \dfrac{9 \times 8 \times 7}{3 \times 2 \times 1} = 84$ different milkshakes

48. Using C's, write out the first five terms of $(x + y)^{57}$.

$(x + y)^{57} = C(57, 0)x^{57} + C(57, 1)x^{56}y + C(57, 2)x^{55}y^2 + C(57, 3)x^{54}y^3$
$\qquad + C(57, 4)x^{53}y^4$

The Complete Solutions and Answers

50. The cost (c) of a trip varies directly as the number of miles (m) driven.
$$c = km$$

52. Factor $x^2 + 9x + 20$
   Look for two things that add to +9 and multiply to 20
   $$= (x + 4)(x + 5)$$

53. Find the largest value that $f(x, y) = 3x + 4y$ becomes subject to the constraints: $x \geq 0$, $y \geq 0$, $3x + 5y \leq 45$, $6x + 5y \leq 60$

$x \geq 0$ and $y \geq 0$ means that the graph is in the first quadrant (Q I).
We did graphing of inequalities in two variables in Chapter 6.
To graph $3x + 5y \leq 45$, you first graph $3x + 5y = 45$. Graphing $3x + 5y = 45$ we did in Chapter 3½ as a review of what was done in beginning algebra. After graphing $3x + 5y = 45$ as a solid line (since there is a $\leq$ and not a $<$ in the original inequality) we test both sides of $3x + 5y = 45$ to see which side to shade in.
Graphing $3x + 5y = 45$ by point-plotting. If $x = 0$, then $y = 9$. (0, 9)
If $y = 0$, then $x = 15$. (15, 0)
Testing (0, 0) in $3x + 5y \stackrel{?}{\leq} 45$. Yes. Shade that region.
Testing (50, 50) in $3x + 5y \stackrel{?}{\leq} 45$. No. Do not shade that region.

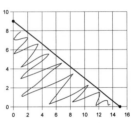

Repeating for $6x + 5y \leq 60$. If $x = 0$, then $y = 12$. (0, 12)
If $y = 0$, then $x = 10$. (10, 0). Solid line since the original inequality was $\leq$.
Test (1, 1) in $6x + 5y \stackrel{?}{\leq} 60$. Yes. Test (40, 50) in $6x + 5y \stackrel{?}{\leq} 60$. No.

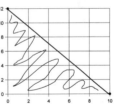

Graph $x \geq 0$, $y \geq 0$, $3x + 5y \leq 45$, $6x + 5y \leq 60$ all on the same graph. The four vertices are at (0, 0), (0, 9), (5, 6), and (10, 0). The (5, 6) was found by solving $3x + 5y = 45$ and $6x + 5y = 60$. (Chapter 5, Systems of Equations). You could use elimination, substitution, or Cramer's rule.

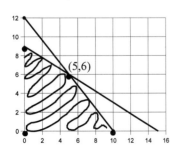

Testing each vertex in $f(x, y) = 3x + 4y$ ...
$f(0, 0) = 0$
$f(0, 9) = 36$
$f(5, 6) = 39$ ⇐ The winner.
$f(10, 0) = 30$

56–60   The Complete Solutions and Answers

56. $\log_{0.5} 16 = -4$ since $0.5^{-4} = (½)^{-4} = 2^4 = 16$
       $\log 0.1 = -1$ $\log 0.1$ means $\log_{10} 0.1$ $10^{-1} = 0.1$
       $\log 1 = 0$ $\log 1$ means $\log_{10} 1$ $10^0 = 1$

59. I need to pick the first place I'll go (4 choices), then pick the second place I'll go (3 choices), then pick the third place I'll go (2 choices), and then pick the last place I'll go (1 choice).

     By the fundamental principle the total number of different ways I'll get to these four spots is 4×3×2×1, which is 24.

     4×3×2×1 can be written as 4! ("four factorial").

60. The hospital is 4 blocks south and 5 blocks east of the milkshake.

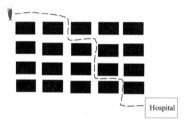

You could describe the route that my mother took as East-East-South-East-South-South-East-South-East.

She had 4 Souths and 5 Easts.

Another route might have been S-S-E-E-E-E-S-E-S.
That would have looked like this:

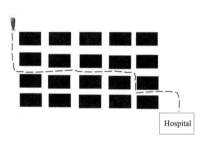

To describe any route, we need 4 Souths and 5 Easts.

     I have 9 slots ___, ___, ___, ___, ___, ___, ___, ___, ___ to fill in with 4 S's and 5 E's.

     Once I figure out where the 4 S's go, the rest are filled in with E's. How many ways are there to put the 4 S's in the 9 slots?

        Easy. C(9, 4)

details: $C(9, 4) = \dfrac{9!}{5!\,4!} = \dfrac{9 \cdot 8 \cdot 7 \cdot 6}{4 \cdot 3 \cdot 2 \cdot 1} = 126$ routes are possible

The Complete Solutions and Answers

62. Solve $\dfrac{6}{2x} = \dfrac{12}{3x+5}$

Cross multiplying
$$6(3x+5) = 2x(12)$$

$\dfrac{a}{b} = \dfrac{c}{d}$ implies $ad = bc$

Distributive property
$$18x + 30 = 2x(12)$$

Arithmetic
$$18x + 30 = 24x$$

Subtract 18x from both sides
$$30 = 6x$$

Divide both sides by 6
$$5 = x$$

65. $(3i)^3 = 3^3 i^3 = 27(i^2)(i) = -27i$

66. Which of these are functions?

Example A: (8, c), (32, d), (40, d) Yes. No element of the domain has two different images.

Example B: (2, r), (5, 5), (5, 3) No. 5 has two different images. It is mapped to 5 and also to 3.

Example C: (1, 2), (2, 3), (3, 4), (4, 5), (5, 6) Yes. No element of the domain has two different images.

Example D: (z, 123), (y, 50), (50, y) Yes. No element of the domain has two different images.

Example E: (v, m), (v, n), (v, p) No. v has more than one image. In fact, v is mapped to three different images: m, n and p.

> The **fast way** to find out whether a set of ordered pairs is a function is to make sure no two ordered pairs have the same first coordinate. If you have (ж, 7) and (ж, rat), it is not a function since a member of the domain, ж, is mapped to two different images.

The definition of function is a rule that assigns to each element of the domain exactly one element in the codomain. In terms of ordered pairs, the definition of a function is a set of ordered pairs in which each first coordinate appears only once.

67–68 The Complete Solutions and Answers

67. (continuing problem #59) Suppose I need to get to the grocery store last because I'm buying ice cream and it will melt if I leave it in the car too long. In how many different orders can I visit those four spots?

I have three choices for the first spot to hit. I then have two choices for the second spot. Then one choice for the third spot. And the grocery store (1 choice) for the last spot.

3×2×1×1 = 6 possible routes to take.

Here are the six ways to visit those four places:

B–OSS–L–GS
B–L–OSS–GS
OSS–B–L–GS
OSS–L–B–GS
L–B–OSS–GS
L–OSS–B–GS

68. Write out the first five terms of $(x + y)^{74}$.

$$(x+y)^{74} = x^{74} + \frac{74}{1!}x^{73}y + \frac{74 \cdot 73}{2!}x^{72}y^2 + \frac{74 \cdot 73 \cdot 72}{3!}x^{71}y^3$$

$$+ \frac{74 \cdot 73 \cdot 72 \cdot 71}{4!}x^{70}y^4 \quad \text{This is so easy. I want to continue for a while.}$$

$$+ \frac{74 \cdot 73 \cdot 72 \cdot 71 \cdot 70}{5!}x^{69}y^5$$

$$+ \frac{74 \cdot 73 \cdot 72 \cdot 71 \cdot 70 \cdot 69}{6!}x^{68}y^6$$

$$+ \frac{74 \cdot 73 \cdot 72 \cdot 71 \cdot 70 \cdot 69 \cdot 68}{7!}x^{67}y^7$$

One thing I like is how the numbers hop-hop-hop on the top of the fraction and land on the x's exponent.

$$+ \frac{74 \cdot 73 \cdot 72 \cdot 71 \cdot 70 \cdot 69 \cdot 68 \cdot 67}{8!}x^{66}y^8$$

I also like it that the sum of the two exponents always equals the original $(x + y)^{74}$.
I also like it that the denominator matches the exponent on the y.

The Complete Solutions and Answers | 71–77

71. $\sqrt{x^2 - x + 16} = 6$

First, isolate the radical.

This has already been done.

Second, square both sides.

Solve.

$$\sqrt{x^2 - x + 16} = 6$$
$$x^2 - x + 16 = 36$$
$$x^2 - x - 20 = 0$$
$$(x - 5)(x + 4) = 0$$
$$x - 5 = 0 \text{ OR } x + 4 = 0$$
$$x = 5 \text{ OR } x = -4$$

Third, check each answer in the original equation.

Checking $x = 5$
$$\sqrt{25 - 5 + 16} \stackrel{?}{=} 6$$
$$\sqrt{36} \stackrel{?}{=} 6 \quad \text{yes}$$

Checking $x = -4$
$$\sqrt{16 - (-4) + 16} \stackrel{?}{=} 6$$
$$\sqrt{36} \stackrel{?}{=} 6 \quad \text{yes}$$

The solution is $x = 5$ and $x = -4$.

74. Let $g: A \to B$ be the function defined by $g(x) = x^2$ and $A = \{3, 5, 9\}$ and $B = \{9, 25, 81\}$.

$g^{-1}(25) = 5$ This is true because $g(5) = 25$.

$g(g^{-1}(81))$
$= g(9)$
$= 81$ 		And $g(g^{-1}(g(g^{-1}(g(5)))))$ equals 25 after you go back and forth between 5 and 25 a bunch of times.

77. Write out what each of these means:

$\sum_{i=1}^{8} 5i = 5 + 5 \cdot 2 + 5 \cdot 3 + 5 \cdot 4 + 5 \cdot 5 + 5 \cdot 6 + 5 \cdot 7 + 5 \cdot 8$

or, if you like, $5 + 5(2) + 5(3) + 5(4) + 5(5) + 5(6) + 5(7) + 5(8)$

$\sum_{i=1}^{4} \log(x + i) = \log(x + 1) + \log(x + 2) + \log(x + 3) + \log(x + 4)$

$\sum_{i=1}^{3} 8x^i y^{i+3} = 8xy^4 + 8x^2 y^5 + 8x^3 y^6$

80–84 The Complete Solutions and Answers

80. The pull (P) that I feel from my mother's carrot cake varies inversely with the distance (d) that I am from it. If I am 10 feet away from it, I feel a pull of 96 pounds drawing me toward it.

When I'm only 4 feet away, I can see it. I can smell it. I can almost reach it. What is the force of attraction at that distance?

Step ①: Find the equation. $P = \dfrac{k}{d}$

Step ②: Find the value of k given 96 pounds of pull at 10 feet.

Substituting that into $P = \dfrac{k}{d}$ we get $96 = \dfrac{k}{10}$

Multiply both sides by 10 $960 = k$

$P = \dfrac{k}{d}$ now becomes $P = \dfrac{960}{d}$

Step ③: Find P when d = 4.

$$P = \dfrac{960}{4}$$

$$P = 240$$

At four feet away, her carrot cake draws me with a force of 240 pounds. Who could resist it?

83. $x^{0.4} x^{0.6} = x^1 = x$
$(y^w)^{3w} = y^{3w^2}$
$zzzzz^4 = z^8$

84. Graph $y = 2x^2 + 5$ from $x = -2$ to $x = 2$.

If x = –2, then $y = 2(-2)^2 + 5 = 13$ The point is (–2, 13).
If x = – 1, then $y = 2(-1)^2 + 5 = 7$ The point is (–1, 7).
If x = 0, then y = 5   The point is (0, 5).
If x = 1, then y = 7   The point is (1, 7).
If x = 2, then y = 13   The point is (2, 13).

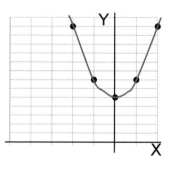

Some people get tired of writing all those If-then statements.

Instead, they make a little chart like this:

x	y
–2	13
–1	7
0	5
1	7
2	13

I think it looks kinda cute.

The Complete Solutions and Answers

85. If h (π) = √2 , then what does h⁻¹(√2) equal?

π $\xrightarrow{h}$ √2   so   √2 $\xrightarrow{h^{-1}}$ π

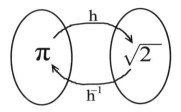

86. Without using a calculator, find the value of $\dfrac{24!}{23!\,4!}$

$$\dfrac{24!}{23!\,4!} = \dfrac{24 \times 23 \times 22 \times 21 \times \ldots \times 3 \times 2 \times 1}{(23 \times 22 \times 21 \times \ldots \times 3 \times 2 \times 1)(4 \times 3 \times 2 \times 1)}$$

$$= \dfrac{24}{4 \times 3 \times 2 \times 1} = 1$$

87. You are the director of the hospital. There are 12 doctors on the staff. You need to assign 5 of them to the emergency ward, 3 to pediatrics, and the remaining 4 will be assigned as lobbyists to Congress.

How many different ways could you make those assignments?

If you are one of the doctors, it doesn't matter whether you are first or last on the list of those assigned to the emergency ward. The order doesn't matter.

There are C(12, 5) ways to choose 5 of the 12 doctors for work in the emergency ward.

After they are chosen, then there are C(7, 3) ways to pick the 3 doctors that will work in pediatrics.

The remaining 4 doctors will be assigned to lobbyist work. That is C(4, 4), which is equal to one. There is only one way to pick a group of 4 out of a group of 4. You pick them all.

By the fundamental principle, the ways for you to make all those assignments is C(12, 5) × C(7, 3) × 1

The details: C(12, 5) × C(7, 3) = $\dfrac{12!}{7!\,5!}$ × $\dfrac{7!}{4!\,3!}$ = (the 7!'s cancel) = $\dfrac{12!}{5!\,4!\,3!}$ = 27,720

89–95 The Complete Solutions and Answers

89. Draw a Venn diagram of the set of all things that contain wood (W) and the set of all furniture (F).

There are things in both W and F (such as a wooden chair) so
W ∩ F ≠ ∅   ∅ is the symbol for the empty set { }.

There are things in W that are not in F (such as a pencil) so it is not the case that W ⊂ F   ⊂ is the symbol for subset. A ⊂ B is true if every element of A is an element of B.

There are things in F that are not in W (such as a metal table) so it is not the case that F ⊂ W.

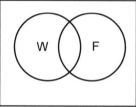

92. $\log_3 9 = 2$ since 3 raised to the second power will equal 9
   $\log_{10} 1000 = 3$ since ten raised to the third power will equal 1,000
   $\log_{17} 17 = 1$ since 17 raised to the first power equals 17.

93. Graph $\dfrac{x}{2} + \dfrac{y}{2\pi} = 1$

Plotting is just approximate.
Just plot (0, 6.3).

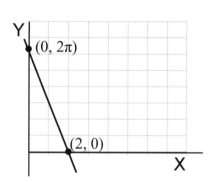

95. Simplify $\dfrac{6x^2 + 7x - 3}{18x - 6}$

$= \dfrac{(2x + 3)(3x - 1)}{6(3x - 1)}$

$= \dfrac{2x + 3}{6}$

Factoring the top:
$6x^2 + 7x - 3$
$= 6x^2 + 9x - 2x - 3$
$= 3x(2x + 3) - (2x + 3)$
$= (2x + 3)(3x - 1)$

Factoring the bottom:
$18x - 6 = 6(3x - 1)$

The Complete Solutions and Answers — 96–99

96. Evaluate $\begin{vmatrix} 5 & 2 & -1 \\ 3 & -3 & 0 \\ 4 & -1 & 0 \end{vmatrix}$ Expanding by the third column

$$-1 \begin{vmatrix} 3 & -3 \\ 4 & -1 \end{vmatrix} - 0 \text{ (its minor)} + 0 \text{ (its minor)}$$

$$= -1(-3 + 12)$$
$$= -9$$

98. Explain whether f is 1-1.

f is not 1-1 since both λ and ∇ have the same image.

Explain whether f is onto.

f is not onto because ℕ is not the image of any element in the domain.

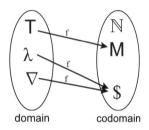

One-to-one means that nothing in the codomain gets "hit" twice.

Onto means that everything in the codomain gets "hit" at least once.

If some function is both 1-1 and onto, that would mean that each element of the codomain gets "hit" exactly once. Functions that are both 1-1 and onto are called **one-to-one correspondences**.

If you can establish a 1-1 correspondence between two sets (a domain and a codomain), then you can say that the sets have the same number of elements. Their cardinalities are equal.

In terms of the diagram, each member of the domain has exactly one arrow leaving it. (f is a function)

No two arrows hit the same target. (f is 1-1)

Every target is hit. (f is onto)

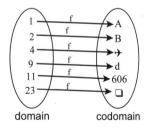

99. Two choices at the bank. Ten choices at the library. Two choices at the office supply store. Four choices at the grocery store.

There are 2×10×2×4 possible ways (= 160) that the story of my visits could have been written.

101–112 The Complete Solutions and Answers

101. Find the median average of 57, 21, 21, 33, 999. The first step is to sort the numbers from smallest to largest: 21, 21, 33, 57, 999. The second step is to pick the one in the middle. The median average is 33.

If there is an even number of numbers that you are averaging, such as 4, 9, 11, 17, 22, 39, then there will be two numbers in the middle. In this case, they are 11 and 17. Add these two numbers up and divide the sum by 2. 11 + 17 equals 28. Twenty-eight divided by 2 equals 14.

Adding two numbers together and dividing by two is called taking the **mean average** of the two numbers. If you want to take the mean average of, say, nine numbers, you add them up and divide the sum by nine. Finding the mean average of a lot of numbers is usually a lot more work than finding the median average.

104. The weight (w) of a rhino varies directly as the cube of its height (h).
$$w = kh^3$$

107. $27^{2/3} = (27^{1/3})^2 = (3)^2 = 9$

$64^{-1/2} = \dfrac{1}{64^{1/2}} = \dfrac{1}{8}$

$(x^{1/2}y^{1/3})^{36} = x^{18}y^{12}$

110. $\log_8 9 - \log_8 3 = \log_8 \dfrac{9}{3} = \log_8 3$

112. Graph $y = 2x + 3$.

There are four steps to graphing *any* equation by point-plotting.
❶ Name any number for the value of x.
❷ Using the equation, find the corresponding y value.
❸ Plot the (x, y) point.
❹ Repeat plotting points (steps ❶–❸) until you can see the shape of the curve. Then connect the dots.

If x = 1, then y = 5. (details: y = 2(1) + 3 = 5) We have the point (1, 5).
If x = 2, then y = 7. (details: y = 2(2) + 3 = 7) We have the point (2, 7).
If x = –1, then y = 1. We have the point (–1, 1).
If x = 0, then y = 3. We have the point (0, 3).

Then we connect the dots.

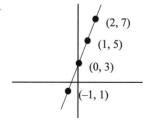

The Complete Solutions and Answers 113–116

113. $(3 + 8i)(2 + 5i) = 6 + 15i + 16i + 40i^2$
$\qquad\qquad\qquad = 6 + 31i - 40$
$\qquad\qquad\qquad = -34 + 31i$

114. For a moment, consider ordered pairs of *numbers*. If you graph two ordered pairs that have the same first coordinate (the same abscissa), what does the graph look like?

Plotting (5, 2) and (5, 9) we notice that one point is directly over the other.

Doing some deductive reasoning . . .

Fact #1: If you express a function as a set of ordered pairs, then no two ordered pairs can have the same first coordinate.

Fact #2: If you graph two ordered pairs that have the same first coordinate, then one point will be directly over the other.

Conclusion: If you look at a graph and notice that two of the points on the graph are vertically aligned (one point over the other), then this is not the graph of a function.

116. If f is defined by $f(x) = x + 7$, what does $f^{-1}(12)$ equal?

$f^{-1}(12)$ asks the question, "What number, call it w, will make $f(w) = 12$ true?"

Since $f(5) = 12$, $f^{-1}(12) = 5$.

Finding inverses sometimes takes more than one step. Suppose g is defined by $g(x) = x^3 - 6$. Then g takes an element of the domain and cubes it and then subtracts 6.

So g^{-1} will have to undo what g does. First it will have to add 6. Then it will take the cube root. Some people who are new to finding inverses might think that the inverse of cube and subtract 6 should be to find the cube root and then add 6. That won't work.

You first put on your socks and then put on your shoes. To undo that, do you first take off your socks? No. That won't work.

Real life example: $g(4) = 4^3 - 6$ which is 58.

People who are new to finding inverses would first take the cube root and then add four. If you do that 58 becomes $\sqrt[3]{58} + 4$. That does not equal the original 4.

Instead, the inverse of $g(4) = 4^3 - 6$ is add 6 and then take the cube root. $g^{-1}(x) = \sqrt[3]{x + 6}$

Trying that, we get $g^{-1}(58) = \sqrt[3]{58 + 6} = \sqrt[3]{64} = 4$, which was what was wanted.

The Complete Solutions and Answers

121. Simplify $\sqrt{6}\sqrt{12}$

$$\sqrt{6}\sqrt{12} = \sqrt{72} = \sqrt{9}\sqrt{8} = \sqrt{9}\sqrt{4}\sqrt{2} = 6\sqrt{2}$$

or using more knowledge or arithmetic

$$\sqrt{6}\sqrt{12} = \sqrt{72} = \sqrt{36}\sqrt{2} = 6\sqrt{2}$$

or using fewer steps

$$\sqrt{6}\sqrt{12} = \sqrt{6}\sqrt{6}\sqrt{2} = 6\sqrt{2}$$

126. $\dfrac{\log_6 144}{\log_6 12} = \log_{12} 144$ (using change of base) $= 2$ (using definition of log)

131. Graph $y = -4x^2 + 7x$ from $x = -5$ to $x = 5$.

Just for fun, I'm going to point-plot eleven points from $x = -5$ to $x = 5$.

x	y
-5	-135
-4	-92
-3	-57
-2	-30
-1	-11
0	0
1	3
2	-2
3	-15
4	-36
5	-65

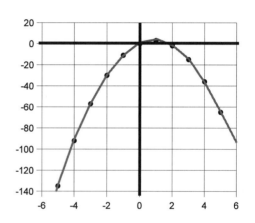

I had to use different scales on the x- and y-axes. Otherwise, the graph would be almost unreadable.

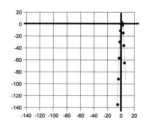

136. What is the distance between $(7, 8)$ and $(-2, 9)$?

$$d = \sqrt{(-2-7)^2 + (9-8)^2} = \sqrt{81 + 1} = \sqrt{82}$$

The Complete Solutions and Answers | 141

141. $29x^3 + 4x^2 + 10x^4 - 7x + 20 \div 2x + 5$

arrange the terms in descending degree

$$2x + 5 \overline{)\, 10x^4 + 29x^3 + 4x^2 - 7x + 20\,}$$

2x into $10x^4$

$$\begin{array}{r} 5x^3 \\ 2x + 5 \overline{)\, 10x^4 + 29x^3 + 4x^2 - 7x + 20\,} \end{array}$$

$5x^3$ times $2x + 5$

$$\begin{array}{r} 5x^3 \\ 2x + 5 \overline{)\, 10x^4 + 29x^3 + 4x^2 - 7x + 20\,} \\ 10x^4 + 25x^3 \end{array}$$

subtract and bring
down the $4x^2$

$$\begin{array}{r} 5x^3 \\ 2x + 5 \overline{)\, 10x^4 + 29x^3 + 4x^2 - 7x + 20\,} \\ \underline{10x^4 + 25x^3} \\ 4x^3 + 4x^2 \end{array}$$

repeat the process
2x into $4x^3$

$$\begin{array}{r} 5x^3 \;\, + 2x^2 \\ 2x + 5 \overline{)\, 10x^4 + 29x^3 + 4x^2 - 7x + 20\,} \\ \underline{10x^4 + 25x^3} \\ 4x^3 + 4x^2 \end{array}$$

$2x^2$ times $2x + 5$

$$\begin{array}{r} 5x^3 \;\, + 2x^2 \\ 2x + 5 \overline{)\, 10x^4 + 29x^3 + 4x^2 - 7x + 20\,} \\ \underline{10x^4 + 25x^3} \\ 4x^3 \;\, + 4x^2 \\ 4x^3 + 10x^2 \end{array}$$

subtract and bring
down the $-7x$

$$\begin{array}{r} 5x^3 \;\, + 2x^2 \\ 2x + 5 \overline{)\, 10x^4 + 29x^3 + 4x^2 - 7x + 20\,} \\ \underline{10x^4 + 25x^3} \\ 4x^3 \;\, + 4x^2 \\ \underline{4x^3 + 10x^2} \\ -6x^2 - 7x \end{array}$$

The Complete Solutions and Answers

repeat the process
2x into −6x²

$$2x + 5 \overline{\smash{\big)}\, \begin{array}{l} 5x^3 + 2x^2 - 3x \\ 10x^4 + 29x^3 + 4x^2 - 7x + 20 \end{array}}$$
$$\underline{10x^4 + 25x^3}$$
$$ 4x^3 + 4x^2$$
$$ \underline{4x^3 + 10x^2}$$
$$ -6x^2 - 7x$$

−3x times 2x + 5

$$2x + 5 \overline{\smash{\big)}\, \begin{array}{l} 5x^3 + 2x^2 - 3x \\ 10x^4 + 29x^3 + 4x^2 - 7x + 20 \end{array}}$$
$$\underline{10x^4 + 25x^3}$$
$$4x^3 + 4x^2$$
$$\underline{4x^3 + 10x^2}$$
$$-6x^2 - 7x$$
$$-6x^2 - 15x$$

subtract and bring
down the +20

$$2x + 5 \overline{\smash{\big)}\, \begin{array}{l} 5x^3 + 2x^2 - 3x \\ 10x^4 + 29x^3 + 4x^2 - 7x + 20 \end{array}}$$
$$\underline{10x^4 + 25x^3}$$
$$4x^3 + 4x^2$$
$$\underline{4x^3 + 10x^2}$$
$$-6x^2 - 7x$$
$$\underline{-6x^2 - 15x}$$
$$8x + 20$$

$$\boxed{-7 - (-15) = -7 + 15 = 8}$$

repeat the process
2x into 8x
4 times 2x + 5
subtract

$$2x + 5 \overline{\smash{\big)}\, \begin{array}{l} 5x^3 + 2x^2 - 3x + 4 \\ 10x^4 + 29x^3 + 4x^2 - 7x + 20 \end{array}}$$
$$\underline{10x^4 + 25x^3}$$
$$4x^3 + 4x^2$$
$$\underline{4x^3 + 10x^2}$$
$$-6x^2 - 7x$$
$$\underline{-6x^2 - 15x}$$
$$8x + 20$$
$$\underline{8x + 20}$$
$$0$$

The Complete Solutions and Answers | 146–166

146. Find the median average of 88, 23, 44, 90. First we sort the numbers from smallest to largest: 23, 44, 88, 90. There are two numbers (44 and 88) which are in the middle.

We find the mean average of 44 and 88.

$$44 + 88 = 132 \qquad 132 \div 2 = 66$$

66 is the median average of 88, 23, 44, 90.

151. Underline the first non-zero digit in:
$\underline{5}3.007$
$\underline{3}000$
$000\underline{4}9.900$
$\underline{6}0.6070$

156. What is the equation of the line with a slope of –4 that passes through the point (2, 9)?

$$m = \frac{y - y_1}{x - x_1} \qquad \text{becomes} \qquad -4 = \frac{y - 9}{x - 2}$$

161. Solve $3x^2 + 5x + 1 = 0$

Perhaps the easiest way to use the quadratic formula is to recite it "minus b, plus-or-minus the square root of b squared minus 4ac, all over 2a" and as you recite it, fill in the numbers.

$$x = \frac{-5 \pm \sqrt{25 - (4)(3)(1)}}{6}$$

$$x = \frac{-5 \pm \sqrt{13}}{6}$$

166.
$$\begin{cases} 6x + 7y - 3z = 44 \\ 2x - 4y + 8z = 39 \\ 5x + 6y + 9z = 82 \end{cases}$$

$y = \dfrac{D_y}{D}$ where D is the determinant formed by the coefficients and D_y is the same as D except that the y coefficients are replaced by the constants.

$$y = \frac{\begin{vmatrix} 6 & 44 & -3 \\ 2 & 39 & 8 \\ 5 & 82 & 9 \end{vmatrix}}{\begin{vmatrix} 6 & 7 & -3 \\ 2 & -4 & 8 \\ 5 & 6 & 9 \end{vmatrix}}$$

| 168–186 | The Complete Solutions and Answers

168. $\sqrt{w+13} - 7 = w$

    Isolate the radical. $\sqrt{w+13} = w + 7$
    Square both sides. $w + 13 = w^2 + 14w + 49$
    Solve. $0 = w^2 + 13w + 36$
         $0 = (w+4)(w+9)$
         $w + 4 = 0$ OR $w + 9 = 0$
         $w = -4$ OR $w = -9$

    Check each answer in the original equation.

Checking $w = -4$
     $\sqrt{-4 + 13} - 7 \stackrel{?}{=} -4$
     $\sqrt{9} - 7 \stackrel{?}{=} -4$
     $3 - 7 \stackrel{?}{=} -4$ yes

Checking $w = -9$
     $\sqrt{-9 + 13} - 7 \stackrel{?}{=} -9$
     $\sqrt{4} - 7 \stackrel{?}{=} -9$
     $2 - 7 \stackrel{?}{=} -9$ no

The solution is $w = -4$.

171. $5i + 7i - 8i = 4i$
      $i^3 = i \cdot i \cdot i = (-1)i = -i$
      $(4i)^2 = 16i^2 = -16$

176. Double <u>underline</u> the gratuitous zeros in:
     5000   There are no extra zeros. 5000 isn't the same as 5.
     30.7<u>0</u>
     239.0809
     298.<u>00</u>   These are gratuitous. 298.00 = 298.

181. $\log_{44} 44^{0.73} = 0.73$ since $44^{0.73} = 44^{0.73}$
      $\log_{6.0735} 1 = 0$ since $6.0735^0 = 1$
      $\log 1000 = 3$ "log" without a base means "$\log_{10}$"

186. $ax^2 + bx + cy^2 + dy = e$

If $b \neq 0$, that means that you will have to complete the square, and that, in turn, means that the center of the ellipse will not be at the origin.
If $d \neq 0$, ditto.

The Complete Solutions and Answers

191. Graph $y - \frac{2}{3}x > 5$

First graph the equality $y - \frac{2}{3}x = 5$

Put it into the $y = mx + b$ form $\quad y = \frac{2}{3}x + 5$

Use a dashed line because the original inequality was > and not ≥.

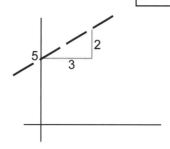

Testing (0, 8), which is above the line, in the original inequality $\quad 8 - \frac{2}{3}(0) \overset{?}{>} 5 \quad$ True.

Testing (0,0), which is below the line, in the original inequality $\quad 0 - \frac{2}{3}(0) \overset{?}{>} 5 \quad$ False.

Shade above the line.

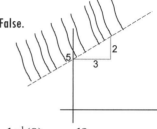

192. If h is defined as $h(x) = (x + 4)^3$, what does $h^{-1}(8)$ equal?

The function h first adds 4 and then cubes the answer.
The inverse function h^{-1} takes the cube root first and then subtracts 4. In symbols: $h^{-1}(x) = \sqrt[3]{x} - 4$
$$h^{-1}(8) = \sqrt[3]{8} - 4 = 2 - 4 = -2$$

Let's check our answer, just for fun. $h(-2) = (-2 + 4)^3 = 2^3 = 8$. Yes.

196. On the first day of the month, I spent $15 on bear food. On the second day, I spent $25. On the third day, $35. Each day I spent $10 more than the previous day. (The bears got hungrier and hungrier.)

How much did I spend in the first 30 days?

This is a two-step problem. First find ℓ.
$a = 15 \quad d = 10 \quad n = 30$
$$\ell = a + (n-1)d = 15 + (29)(10) = 305$$

Then find s.
$$s = \frac{n}{2}(a + \ell) = 15(15 + 305) = \$4{,}800.$$

$4,800 does not include the hospital bills from feeding bears.

201–211 | The Complete Solutions and Answers

201. Find the mean and the median average of 7, 8, 10, 16, 14.

To find the mean average, add up the five numbers and divide the sum by five. $7 + 8 + 10 + 16 + 14 = 55 \qquad 55 \div 5 = 11$

11 is the mean average.

To find the median average, sort the numbers from smallest to largest—7, 8, 10, 14, 16—and pick the one in the middle.

The median average is 10.

206. $\dfrac{5}{x+2} + \dfrac{3}{x-7} = \dfrac{5(x-7)}{(x+2)(x-7)} + \dfrac{3(x+2)}{(x-7)(x+2)}$

$= \dfrac{5x - 35 + 3x + 6}{(x+2)(x-7)}$

$= \dfrac{8x - 29}{(x+2)(x-7)}$

211. Solve $\dfrac{x}{x+2} = \dfrac{6}{x^2 + x - 2}$ Hint: Solve as a fractional equation rather than cross-multiplying. If you cross-multiply, you will get a cubic equation (an equation containing x³).

$\dfrac{x}{x+2} = \dfrac{6}{(x-1)(x+2)}$ Factor the denominator(s)

$\dfrac{x(x-1)(x+2)}{x+2} = \dfrac{6(x-1)(x+2)}{(x-1)(x+2)}$ Multiply by (x – 1)(x + 2), which is an expression that both denominators evenly divide into

$x(x-1) = 6$

$x^2 - x = 6$

$x^2 - x - 6 = 0$ Solving by factoring

$(x-3)(x+2) = 0$

$x - 3 = 0 \quad \text{OR} \quad x + 2 = 0$

$x = 3 \quad \text{OR} \quad x = -2$

Checking each possible answer in the original equation:

x = 3

$\dfrac{3}{3+2} \stackrel{?}{=} \dfrac{6}{9+3-2} \qquad \dfrac{3}{5} \stackrel{?}{=} \dfrac{6}{10}$ yes

x = –2

$\dfrac{-2}{-2+2}$ Stop right there. We have division by zero. no

The final answer is x = 3.

The Complete Solutions and Answers 216–226

216. Solve $(2x + 5)(3x - 1) + 5 = 0$

$6x^2 - 2x + 15x - 5 + 5 = 0$

$6x^2 + 13x = 0$ This can be solved by factoring.

$x(6x + 13) = 0$

$x = 0$ OR $6x + 13 = 0$

$x = 0$ OR $x = \dfrac{-13}{6}$

221. Solve $\begin{cases} 4x - 3y = 10 \\ 7x + 2y = 3 \end{cases}$

Multiply the first equation by 2
and the second equation by 3
$\begin{cases} 8x - 6y = 20 \\ 21x + 6y = 9 \end{cases}$

Add to eliminate the y terms

$29x = 29$

$x = 1$

Back substitute into the first equation

$4(1) - 3y = 10$

$-6 = 3y$

$-2 = y$

226. Solve by graphing $\begin{cases} 3x + y = 5 \\ 7x + 6y = -3 \end{cases}$

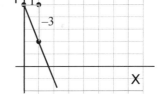

Some people like to plot $3x + y = 5$ by point-plotting. They let $x = 0$ and y is 5. $(0, 5)$
They let $x = 2$ and y is -1. $(2, -1)$
I prefer the point-slope form, $y = mx + b$. $3x + y = 5$ becomes $y = -3x + 5$ and I can slash in the graph:

Similarly, put $7x + 6y = -3$ into $y = mx + b$ form.
$y = -\dfrac{7}{6}x - \dfrac{1}{2}$ and almost instantly draw that graph:

Putting both of these on the
the same set of axes:

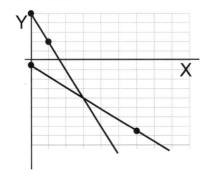

The point of intersection is about $(3, -4)$.

231–240 | The Complete Solutions and Answers

231. Place into standard form for an ellipse

$$49x^2 - 196x + y^2 + 6y + 156 = 0$$

$$49(x^2 - 4x) + y^2 + 6y = -156$$ 　In order to complete the square, the coefficient of the squared term must be equal to one.

$$49(x^2 - 4x + 4) + y^2 + 6y + 9 = -156 + 196 + 9$$ 　Adding +4 inside the parentheses, adds 196 to the left side of the equation.

$$49(x - 2)^2 + (y + 3)^2 = 49$$

$$\frac{(x-2)^2}{1} + \frac{(y-(-3))^2}{7^2} = 1$$ 　This is an ellipse centered at (2, −3) with a vertical semi-major axis with length equal to 7 and a horizontal semi-minor axis with length equal to 1.

236. Find the equation of the circle the circumscribes the ellipse

$$\frac{(x-3)^2}{36} + \frac{(y-5)^2}{64} = 1$$

The circle will have the same center as the ellipse.
The center of the ellipse is (3, 5).
The circle will have its radius equal to the length of the semi-major axis, which is 8.
The equation of the circle is $(x - 3)^2 + (y - 5)^2 = 64$

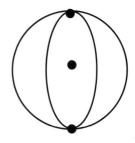

240. Is this a function?

　　　　　Nebraska → moon
　　　　　Kansas → purple
　　　　　South Dakota → #
　　　　　New Jersey → 8
　　　　　Kansas → coat

Kansas has two images.　It is not a function.

　　If it is going to be a function, each element in the domain must have exactly one image.

100

The Complete Solutions and Answers

245. How long (L) I have to wait in line at the amusement park varies directly as the square root of the number of kids (n) at the park.

If there are 20 kids, I have to wait 7 minutes in line. How long is the wait if there are 180 kids?

    Step ①: Find the equation. $L = k\sqrt{n}$

    Step ②: Find the value of k.   If there are 20 kids, I wait 7 minutes.
    Substitute that into $L = k\sqrt{n}$ $7 = k\sqrt{20}$

    Divide both sides by $\sqrt{20}$ $\dfrac{7}{\sqrt{20}} = k$

    $L = k\sqrt{n}$ now becomes $L = \dfrac{7\sqrt{n}}{\sqrt{20}}$

    Step ③: Find L when n = 180. $L = \dfrac{7\sqrt{180}}{\sqrt{20}}$

    Since $\dfrac{\sqrt{a}}{\sqrt{b}} = \sqrt{\dfrac{a}{b}}$ $L = 7\sqrt{9} = 7(3) = 21$

With 180 kids, I will have to wait 21 minutes in line.

246. $\dfrac{a + \sqrt{c}}{a - \sqrt{c}} = \dfrac{a + \sqrt{c}}{a - \sqrt{c}} \cdot \dfrac{a + \sqrt{c}}{a + \sqrt{c}} = \dfrac{(a + \sqrt{c})^2}{a^2 - c}$

251. Place $\dfrac{7}{4-i}$ in the form a + bi.

$\dfrac{7}{4-i} = \dfrac{7}{4-i} \cdot \dfrac{4+i}{4+i} = \dfrac{28 + 7i}{16 - i^2} = \dfrac{28 + 7i}{17} = \dfrac{28}{17} + \dfrac{7}{17}i$

256. Solve $\begin{cases} 5x + 2y = 41 \\ 3x - 2y = 15 \end{cases}$

If we add the two equations together, the y terms will disappear. That is why it is called the elimination method.

Adding the two equations: $8x = 56$ ➡ $x = 7$

To find the corresponding value of y, we substitute the x = 7 into any equation containing x and y.

x = 7 into the first equation: $5(7) + 2y = 41$ ➡ $35 + 2y = 41$ ➡ y = 3

If I substituted into the second equation, I would get the same answer.

                 $3(7) - 2y = 15$ ➡ $21 - 2y = 15$ ➡ y = 3

| 261–271 | The Complete Solutions and Answers

261. Place $\dfrac{4-5i}{2+3i}$ in the form $a+bi$

$$\dfrac{4-5i}{2+3i} = \dfrac{4-5i}{2+3i} \cdot \dfrac{\mathbf{2-3i}}{\mathbf{2-3i}}$$

$$= \dfrac{8-12i-10i+15i^2}{4-9i^2}$$

$$= \dfrac{8-22i-15}{4+9}$$

$$= \dfrac{-7-22i}{13}$$

$$= -\dfrac{7}{13} - \dfrac{22}{13}i$$

266. Every year my book collection increases by 6%. How many years (to the nearest year) will it take to double my collection?

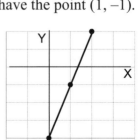

$(1.06)^x = 2$

$\log(1.06)^x = \log 2$

$x \log 1.06 = \log 2$

$x = \dfrac{\log 2}{\log 1.06}$

$x \approx \dfrac{0.30102999566398119521373889472449}{0.025305865264770240846731186351 75}$ Just for fun I used my big calculator.

$x \approx 11.89566104594188560828201787603 2 \doteq 12$ years

271. Graph $y - 3x = -4$

It will be easier if we add $3x$ to both sides of the equation: $y = 3x - 4$

If $x = 0$, then $y = -4$. (details: $y = 3(0) - 4 = -4$)   We have the point $(0, -4)$.
If $x = 1$, then $y = -1$. (details: $y = 3(1) - 4 = -1$)   We have the point $(1, -1)$.
If $x = 2$, then $y = 2$.   We have the point $(2, 2)$.
You could let $x = 59.3907$, then y would equal 174.1721, but that would be silly (and a lot of unnecessary work).

The Complete Solutions and Answers

273. Find a value for x so that the median average of 6, x, 10 is less than the mean average of those three numbers. This may take a little experimenting.

If x is equal to 8, then both the mean average and the median average of 6, 8, 10 are equal to 8.

If x is 7, then the median average is 7. The mean average is
6 + 7 + 10 = 23   23 ÷ 3 = 7⅔ This works. The median average is less than the mean average.

One guess at this point, would be that any x < 8, will work.

Can this question be answered using algebra?

Case 1: 6 < x < 10

The median average of 6, x, 10 will always be x.

The mean average of 6, x, 10 is (6 + x + 10)/3, which is 16/3 + x/3.

We want the median average to be less than the mean.

We want x to be less than 16/3 + x/3.

$$x < 16/3 + x/3$$

Using the methods of solving inequalities (Chapter 12 of *Life of Fred: Beginning Algebra Expanded Edition*), we multiply through by 3:

$$3x < 16 + x$$

Subtract x from both sides:

$$2x < 16$$

Divide both sides by 2:

$$x < 8 \quad \text{which is what we wanted.}$$

Case 2: x < 6 < 10

In this case, the median average will be 6.

The mean average will still be 16/3 + x/3.

We want the median average to be less than the mean average.

$$6 < 16/3 + x/3$$

Multiply through by 3:   18 < 16 + x

Subtract 16 from both sides: 2 < x

Surprisingly, when x is between 2 and 6, it will work: the median will be less than the mean.

Case 3: x = 6   The median (6) will be less than the mean (7⅓).

Case 4: 6 < 10 < x   The median is 10. The mean is still 16/3 + x/3.

10 < 16/3 + x/3   Multiply by 3: 30 < 16 + x

Subtract 16:   14 < x   So when x is greater than 14 it will work.

This problem had a surprising answer: 2 < x < 8 and 14 < x.

The Complete Solutions and Answers

275. Solve $\sqrt{y^2 - 24} = 5$

Isolate the radical.
This has already been done.
Square both sides.

$$\sqrt{y^2 - 24} = 5$$
$$y^2 - 24 = 25$$
$$y^2 = 49$$
$$y = \pm 7$$

Check each answer in the original equation.

checking $y = 7$
$$\sqrt{49 - 24} \stackrel{?}{=} 5$$
$$\sqrt{25} \stackrel{?}{=} 5 \quad \text{yes}$$

checking $y = -7$
$$\sqrt{49 - 24} \stackrel{?}{=} 5$$
$$\sqrt{25} \stackrel{?}{=} 5 \quad \text{yes}$$

Both $y = 7$ and $y = -7$ work. The solution is $y = \pm 7$.

277. $\dfrac{1}{x+5} + \dfrac{2x+11}{x^2 + 9x + 20}$

$= \dfrac{1}{x+5} + \dfrac{2x+11}{(x+4)(x+5)}$ factor denominator

$= \dfrac{(x+4)}{(x+5)(x+4)} + \dfrac{2x+11}{(x+4)(x+5)}$ interior decoration

$= \dfrac{x + 4 + 2x + 11}{(x+5)(x+4)}$

$= \dfrac{3x + 15}{(x+5)(x+4)} = \dfrac{3(x+5)}{(x+5)(x+4)} = \dfrac{3}{x+4}$

simplifying the answer

279. Solve $\begin{cases} 7x + 3y = 33 \\ 9x - 6y = 3 \end{cases}$

I have a choice. ❶ Multiply the first equation by 2 and add the equations. That will eliminate the y term, or ❷ Multiply the first equation by 9 and the second equation by −7 and add the equations. That will eliminate the x term. I am not crazy. The first choice ❶ is much less work.

$\begin{cases} 14x + 6y = 66 \\ 9x - 6y = 3 \end{cases}$ adding the equations ⟹ $23x = 69$ ⟹ $x = 3$

Substituting $x = 3$ into first equation. $7(3) + 3y = 33$ ⟹ $3y = 12$ ⟹ $y = 4$

The Complete Solutions and Answers 281–289

281. $x^3 x^5 = x^8$
$y^4 y^2 y^3 = y^9$
$z^6/z^3 = z^3$

283. $\sqrt{x+44} = 7$

There are three steps in solving radical equations.   (Radical equations are equations that have the unknown under a radical sign $\sqrt{}$.)

First: Isolate the radical on one side of the equation.
   So, for example, $\sqrt{y} + 8 = 10$ should be turned into $\sqrt{y} = 2$
Second: Square both sides of the equation.
Third: Check your answer in the original problem.   (Sometimes squaring introduces answers that don't work in the original problem. They are called extraneous roots.)

Step 1 is already done.
The radical is already by itself
on one side of the equation. $\sqrt{x+44} = 7$
Step 2: Square both sides. $x + 44 = 49$
The square of a square root is easy. $(\sqrt{m})^2 = m$.
   Subtract 44 from both sides $x = 5$

Step 3: Check the answer in the original problem.
$\sqrt{5+44} \stackrel{?}{=} 7$
$\sqrt{49} \stackrel{?}{=} 7$ Yes. It checks. $x = 5$ is the solution.

285. $(7+3i)(2-3i) = 14 - 21i + 6i - 9i^2 = 14 - 15i + 9 = 23 - 15i$

287. What is the equation of the line with a slope of $\frac{3}{4}$ that passes through the point $(-5, -1)$?

$m = \frac{y - y_1}{x - x_1}$ becomes $\frac{3}{4} = \frac{y - (-1)}{x - (-5)}$ OR $\frac{3}{4} = \frac{y+1}{x+5}$

289. What is the equation of the line that passes through $(8, 9)$ and $(2, 5)$?

$\frac{y - y_1}{x - x_1} = \frac{y_2 - y_1}{x_2 - x_1}$ becomes $\frac{y - 9}{x - 8} = \frac{5 - 9}{2 - 8}$

OR $\frac{y-9}{x-8} = \frac{2}{3}$ OR $3y - 2x = 11$

290–296 The Complete Solutions and Answers

290. What is the 503rd term of $10 + 18 + 26 + 34 + 42 + \ldots$?

$\ell = a + (n-1)d = 10 + (502)8 = 10 + 4016 = 4026$

291. What is the equation of the line that passes through $(32, -4)$ and $(30, 9)$?

$$\frac{y - y_1}{x - x_1} = \frac{y_2 - y_1}{x_2 - x_1} \quad \text{becomes} \quad \frac{y + 4}{x - 32} = \frac{9 + 4}{30 - 32}$$

$$\text{OR} \quad \frac{y + 4}{x - 32} = \frac{13}{-2} \quad \text{OR} \quad 13x + 2y = 408$$

293.
$4x(3x + 5y) = 12x^2 + 20xy$

$-3y(6y^2 - 2y) = -18y^3 + 6y^2$

$7x^2y(3xy + 8x^4) = 21x^3y^2 + 56x^6y$

295. The cost (c) of cleaning up junk varies directly as the weight (w). If it cost $80 to clean up 30 pounds of junk, how much would it cost to clean up 50 pounds?

Step ①: Find the equation. $c = kw$

Step ②: Find the value of k. It cost $80 to clean up 30 pounds.

Substituting that into $c = kw$, we get $80 = k(30)$

Divide both sides by 30 $\frac{80}{30} = k$

Simplifying $\frac{8}{3} = k$

$c = kw$ now becomes $c = \frac{8}{3}w$

Step ③: Find c when $w = 50$ $c = \frac{8}{3}(50)$

Doing the arithmetic $c = 133\frac{1}{3}$

Cleaning up 50 pounds would cost $133.33 (after rounding).

296. Factor
$x^2 + 20x + 100 = (x + 10)(x + 10)$ or $(x + 10)^2$
$y^2 + 11y + 18 = (y + 2)(y + 9)$
$z^2 + 6z + 9 = (z + 3)(z + 3)$ or $(z + 3)^2$
$3x^2 + 27x + 54$ *Always look for a common factor first.*
$= 3(x^2 + 9x + 18)$
$= 3(x + 3)(x + 6)$

The Complete Solutions and Answers

297. Graph $x = 3y - 1$
If $y = 1$, then $x = 2$. (details: $x = 3(1) - 1 = 2$) We have the point (2, 1).
If $y = 2$, then $x = 5$. We have the point (5, 2). Not (2, 5)!
If $y = 0$, then $x = -1$. We have the point (-1, 0).

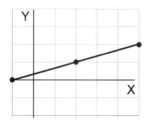

299. Given the coordinates of A are (2, 6).
The coordinates of B are (8, 14).

❶ Find the coordinates of C.
B and C have the same x-coordinate.
So C is of the form (8, ?).

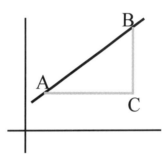

A and C have the same y-coordinate.
So C has coordinates (8, 6).

❷ Find the length of BC.
The length of BC is the distance between (8, 14) and (8, 6).
The distance is 8. (details: 14 – 6) (Distances are never negative.)

❸ Find the length of AC.
The length of AC is the distance between (2, 6) and (8, 6). It is 6.

❹ Find the length of AB.
By the Pythagorean theorem, $6^2 + 8^2 = c^2$

$$36 + 64 = c^2$$
$$100 = c^2$$
$$10 = c$$

Normally, in algebra we would write $c = \pm 10$, but lengths are never negative.

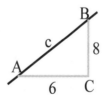

The length of AB is 10.

300 | The Complete Solutions and Answers

300. The domain and the codomain are the natural numbers $\{1, 2, 3, 4, 5, 6, 7, \ldots\}$. g is defined by the rule $g(x) = x^2$.

Is g a function? Is g 1-1? Is g onto?

Under the rule $g(x) = x^2$: $1 \xrightarrow{g} 1$
$2 \xrightarrow{g} 4$
$3 \xrightarrow{g} 9$
$4 \xrightarrow{g} 16$ etc.

Every member of the domain has exactly one image in the codomain. Therefore, g is a function.

Can two different elements of the domain have the same image? Could, for example, 4^2 and 395^2 both have the same answer? No. The function g is 1-1.

Is every member of the codomain an image of some element of the domain? What natural number n, when squared, will equal 5? There isn't one. The function g is not onto.

301. My dog Wufwuf has 4 fleas on the day I have given her a bath. We'll call that day #1. On the next day (day #2) she has 28 fleas and the day after that (day #3), 196 fleas. The sequence is $4, 4 \cdot 7, 4 \cdot 7^2, 4 \cdot 7^3, \ldots$.

(4·7 means 4 times 7)

How many fleas does Wufwuf have on day #22?
$a = 4, r = 7, n = 22 \quad \ell = ar^{n-1} = 4 \cdot 7^{21}$

I, your reader, have a question. I don't want to work on my hand calculator for ten minutes, multiplying 7 times itself. Can you tell me what 4×7^{21} is equal to?

Sure. It's 2,234,183,456,333,136,028 fleas. Two quintillion, two hundred thirty-four quadrillion, one hundred eighty-three trillion, four hundred fifty-six billion, three hundred thirty-three million, one hundred thirty-six thousand, twenty-eight fleas.

Wait! How did you do that!

I told you I have large calculator. If you do it on a regular hand-held calculator, you will get an answer of 2.2342×10^{18}.

I do have a regular hand-held calculator. Just tell me how you can get the 7^{21} part. I can multiply by 4. I mean without having to hit × 7 twenty-one times.

(continued next page)

The Complete Solutions and Answers

On regular scientific calculators (that have sin, cos, log, ln buttons) there is usually a button that says either x^y or y^x. If you want 7^{21}, you can usually do this by punching 7, then x^y, then 21, then =.

Now that I have your attention, I've got one more tiny question.

Fire away. I'm all ears.

Inside of the calculator, how does it go from punching 7, then x^y, then 21, then = to finding the answer. I'm pretty sure that there isn't a little man inside the calculator that computes 7×7.

You are right.

Okay. Tell me how does my calculator do x^y?

Here is my best guess. (You are going to need to remember some stuff from Chapter 3 on logs.)

To compute 7^{21}, your calculator first computes   ln 7

(Recall ln 7 means $\log_e 7$. The natural log of 7. I like to use ln instead of log, because ln is what is needed throughout calculus. In contrast, log 7 means $\log_{10} 7$.)

It will get an answer like	1.9459101
Then it multiplies by 21 and gets	40.864113
Then it takes the antilog and gets	5.5855×10^{17}

In math, here's why that works: take 7   find ln ⟹ ln 7   multiply by 21 ⟹ 21(ln 7)   which by the birdie rule is equal to $\ln 7^{21}$ take antilog ⟹ 7^{21}

Three steps—ln, multiply by 21, antilog—that's short. Wait a minute! How in the world does my calculator find ln 7? There isn't a little man inside that has all the values of the natural logarithm memorized.

No little man inside. When you punch in 7 and hit the ln key, your calculator computes what ln 7 should be. It's a simple computation.

$$\ln 7 = \frac{6}{7} + \frac{1}{2}(\frac{6}{7})^2 + \frac{1}{3}(\frac{6}{7})^3 + \frac{1}{4}(\frac{6}{7})^4 + \ldots$$

You can't kid me. That series goes on forever. The calculator will be computing till Kingdom come.

No it won't. Each term is smaller than the previous one.
$\ln 7 \approx 0.85 + 0.37 + 0.21 + 0.13 + \ldots$

(continued next page)

The Complete Solutions and Answers

It just computes until the next term doesn't add that much to the final answer. Then it stops and spits out the answer.

The series for ln x is $= (\frac{x-1}{x}) + \frac{1}{2}(\frac{x-1}{x})^2 + \frac{1}{3}(\frac{x-1}{x})^3 + \ldots$

That's simple enough.
So ln 5 is $(\frac{5-1}{5}) + \frac{1}{2}(\frac{5-1}{5})^2 + \frac{1}{3}(\frac{5-1}{5})^3 + \ldots$

But I have to admit, I don't like ln (the natural log or log_e). I like the common log (log_{10}) instead. My last question: What's the series for log x?

There isn't a nice series for log_{10} x.

Then what does my calculator do when I punch the log button?

If you want log 4.56, for example, my guess is that it computes ln 4.56 instead.

It finds $(\frac{4.56-1}{4.56}) + \frac{1}{2}(\frac{4.56-1}{4.56})^2 + \frac{1}{3}(\frac{4.56-1}{4.56})^3 + \ldots$

Big deal. I want log_{10} 4.56, not log_e 4.56. What good is ln 4.56?

It is a lot of good. Once you have ln 4.56, it takes one step to get log 4.56. You just multiply by 0.43429448190325182765112891891661. The calculator has memorized this one constant. That's all you need. Watch: $0.43429448190325182765112891891661 \times \ln 4.56 = \log_{10} 4.56$

What kind of magic is this? Why didn't you ever explain this to me before?

I did. This is called the change of base rule (found at the top of page 24 in this book): $(\log_c b)(\log_b a) = \log_c a$
In the case we are talking about: $(\log_{10} e)(\log_e 4.56) = \log_{10} 4.56$
where $\log_{10} e$ equals 0.43429448190325182765112891891661.

But you never told me how you found out that the series for ln x is equal to $(\frac{x-1}{x}) + \frac{1}{2}(\frac{x-1}{x})^2 + \frac{1}{3}(\frac{x-1}{x})^3 + \ldots$ **. This is my last question. Why is that true?**

Your *last* question? You said that a half page ago.

Did I? (Another question!)

The gateway for showing ln x is $= (\frac{x-1}{x}) + \frac{1}{2}(\frac{x-1}{x})^2 + \ldots$ will be when we talk about power series in the second year of calculus. I have to save something for later. ☺

The Complete Solutions and Answers

303. $\dfrac{7}{\sqrt{x+6}} = \dfrac{7}{\sqrt{x+6}} \cdot \dfrac{\sqrt{x+6}}{\sqrt{x+6}} = \dfrac{7\sqrt{x+6}}{x+6}$

305. Every year the administration at KITTENS University changes its plans for the new Dean's Office.

   When the building was first designed, it was 1,200 square feet.
   Each year the plans increase its square footage by 7%.
   How many years before the Dean's Office would be 15,000 square feet?

$$1200(1.07)^x = 15000$$

Take the log of both sides	$\log 1200(1.07)^x = \log 15000$
Product Rule	$\log 1200 + \log (1.07)^x = \log 15000$
Birdie Rule	$\log 1200 + x \log 1.07 = \log 15000$
Algebra	$x = \dfrac{\log 15000 - \log 1200}{\log 1.07}$
Using a calculator to approximate the answer	$x \approx \dfrac{4.176 - 3.079}{0.0294} \doteq 37$ years

307. Simplify

$$\dfrac{x^2 - y^2}{3x^2 + 5xy + 2y^2}$$

$$= \dfrac{(x+y)(x-y)}{(3x+2y)(x+y)}$$

$$= \dfrac{x-y}{3x+2y}$$

Factor top:
$$x^2 - y^2 = (x+y)(x-y)$$
Factor bottom:
$$3x^2 + 5xy + 2y^2$$
$$= 3x^2 + 2xy + 3xy + 2y^2$$
$$= x(3x+2y) + y(3x+2y)$$
$$= (3x+2y)(x+y)$$

309. Solve $(2x^2 + 3)(3x - 1) = 6x(x^2 + 3)$

$$6x^3 - 2x^2 + 9x - 3 = 6x^3 + 18x$$

$$0 = 2x^2 + 9x + 3 \quad \text{It doesn't factor. Use the quadratic formula.}$$

$$x = \dfrac{-9 \pm \sqrt{81 - (4)(2)(3)}}{4} = \dfrac{-9 \pm \sqrt{57}}{4}$$

| 311–317 | The Complete Solutions and Answers |

311. The number of drops of sweat (s) varies directly as the number of miles (m) that you walk. If you walk 12 miles, you will generate 5 drops of sweat. How many drops of sweat is associated with walking 15 miles?

Step ①: Find the equation. $s = km$

Step ②: Find the value of k.   5 drops for 12 miles

Substituting that into $s = km$, we get $5 = k(12)$

Divide both sides by 12 $\frac{5}{12} = k$

$s = km$ now becomes $s = \frac{5}{12} m$

Step ③: Find s when m = 15 $s = \frac{5}{12}(15)$

Doing the arithmetic $s = \frac{5}{\underset{4}{12}}(\overset{5}{15}) = \frac{25}{4} = 6¼$

Walking 15 miles would generate 6¼ drops of sweat.

313. $i^{1,000,000} = (i^4)^{250,000} = 1^{250,000} = 1$ Recall $(x^m)^n = x^{mn}$

315. $\frac{5}{x-6} + \frac{-3x-18}{x^2-36}$

$= \frac{5}{x-6} + \frac{-3x-18}{(x-6)(x+6)}$ factor denominator(s)

$= \frac{5(x+6)}{(x-6)(x+6)} + \frac{-3x-18}{(x-6)(x+6)}$ interior decoration

$= \frac{5x+30-3x-18}{(x-6)(x+6)}$

$= \frac{2x+12}{(x-6)(x+6)}$ and now simplify the fraction

$= \frac{2(x+6)}{(x-6)(x+6)} = \frac{2}{x-6}$

317. $x^2 = 49$
$x = \pm 7$ The hardest part of solving pure quadratics is remembering to put in the $\pm$.

$y^2 = 10^6$
$y = \pm\sqrt{10^6} = \pm(10^6)^{½} = \pm 10^3$ or $\pm 1,000$

$z^2 = 33$
$z = \pm\sqrt{33}$ This cannot be simplified any further.

The Complete Solutions and Answers — 318

318. $\dfrac{8x - 29}{(x + 2)(x - 7)} = \dfrac{A}{x + 2} + \dfrac{B}{x - 7}$

We are trying to find values for A and B so that this equation will be true for all values of x.

Just as we did in Chapter 4½ in solving fractional equations, we eliminate the denominators. We multiply each term by an expression that each denominator will divide evenly into. This will be the first step in working with each of the four cases of partial fractions.

Multiply each term by $(x + 2)(x - 7)$

$$\dfrac{(8x - 29)(x + 2)(x - 7)}{(x + 2)(x - 7)} = \dfrac{A(x + 2)(x - 7)}{x + 2} + \dfrac{B(x + 2)(x - 7)}{x - 7}$$

All the denominators disappear.

$$8x - 29 = A(x - 7) + B(x + 2)$$

In case 1 (distinct linear factors in the denominator) there is a neat shortcut that cuts the work down. Our goal is to find the values of A and B. They are to make the equation true for every value of x. We get to choose any value of x that we like.

Let $x = 7$ $\quad 8(7) - 29 = B(7 + 2)$ $\quad$ Did you notice that the A disappeared?

$\qquad\qquad\quad 56 - 29 = 9B$

$\qquad\qquad\qquad\quad 3 = B$

Let $x = -2$ $\qquad\qquad\qquad\qquad 8(-2) - 29 = A(-2 - 7)$

$\qquad\qquad\qquad\qquad\qquad\quad -16 - 29 = -9A$

$\qquad\qquad\qquad\qquad\qquad\qquad\quad 5 = A$

$\dfrac{8x - 29}{(x + 2)(x - 7)} = \dfrac{A}{x + 2} + \dfrac{B}{x - 7}$ becomes $\dfrac{5}{x + 2} + \dfrac{3}{x - 7}$

Just for fun, I am going to add $\dfrac{5}{x + 2}$ and $\dfrac{3}{x - 7}$ and see if I get $\dfrac{8x - 29}{(x + 2)(x - 7)}$

$\dfrac{5}{x + 2} + \dfrac{3}{x - 7}$

$= \dfrac{5(x - 7)}{(x + 2)(x - 7)} + \dfrac{3(x + 2)}{(x - 7)(x + 2)}$ $\quad$ In Beginning Algebra, we called this Interior Decorating.

$= \dfrac{5(x - 7) + 3(x + 2)}{(x + 2)(x - 7)}$

$= \dfrac{5x - 35 + 3x + 6}{(x + 2)(x - 7)}$

$= \dfrac{8x - 29}{(x + 2)(x - 7)}$ $\quad$ It checked!

321–325 The Complete Solutions and Answers

321. Solve $x = \sqrt{6-x}$

First, isolate the radical.

This has already been done. $\qquad x = \sqrt{6-x}$

Second, square both sides. $\qquad x^2 = 6 - x$

Solving by factoring $\qquad x^2 + x - 6 = 0$

$\qquad\qquad\qquad (x-2)(x+3) = 0$

$\qquad\qquad\qquad x - 2 = 0 \quad \text{OR} \quad x + 3 = 0$

$\qquad\qquad\qquad x = 2 \quad \text{OR} \quad x = -3$

Third, check each answer in the original equation.

checking $x = 2$

$\qquad\qquad 2 \stackrel{?}{=} \sqrt{6-2}$

$\qquad\qquad 2 \stackrel{?}{=} \sqrt{4}\quad$ yes

checking $x = -3$

$\qquad\qquad -3 \stackrel{?}{=} \sqrt{6-(-3)}\quad$ no

A $\sqrt{}$ can never equal a negative number. $\qquad$ The solution is $x = 2$.

323. Graph $y = \frac{3}{4}x - 2$

The y-intercept is at $(0, -2)$.

The slope is $\frac{3}{4}$

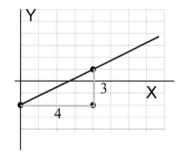

325. Factor $7x^2 + 37x + 10$

First split 37x into two things that add to +37x and multiply to 70x² (= 7x² times 10)

$\qquad = 7x^2 + 35x + 2x + 10$

Then factor by grouping

$\qquad = 7x(x+5) + 2(x+5)$

$\qquad = (x+5)(7x+2)$

Factor $2x^2 + 11x + 12$

First split +11x into two things that add to +11x and multiply to 24x² (= 2x² times 12)

$\qquad = 2x^2 + 3x + 8x + 12$

Then factor by grouping

$\qquad = x(2x+3) + 4(2x+3)$

$\qquad = (2x+3)(x+4)$

The Complete Solutions and Answers

327. I lost my hose so I use water from the tub and from the kitchen sink to water my lawn.

Tub water contains 3 parts of iron and 1 part of sulfur per gallon.

Sink water contains 2 parts of iron and 4 parts of sulfur per gallon.

My lawn needs at least 18 parts of iron and 8 parts of sulfur.

It takes a minute to haul a gallon of tub water out to the lawn. It takes 6 minutes to haul a gallon of sink water out to the lawn. I want to minimize the amount of time used in hauling water.

What is asked in this problem? How many gallons of tub water and how many gallons of sink water should I use?

Let x = number of gallons of tub water I use.
Let y = number of gallons of sink water I use.
Now, reading the English and using the previous two lines, I can say that
$3x + 2y$ is the amount of iron I'll get.
$x + 4y$ is the amount of sulfur I'll get.

And since My lawn needs at least 18 parts of iron and 8 parts of sulfur, $3x + 2y \geq 18$
$$x + 4y \geq 8$$

Since x = number of gallons of tub water I use and it takes a minute to haul a gallon of tub water out to the lawn, I spend x minutes hauling tub water.

Since y = number of gallons of sink water I use and it takes 6 minutes to haul a gallon of sink water out to the lawn, I spend $6y$ minutes hauling sink water.

I spend a total of $x + 6y$ minutes. I want to minimize this.
The objective function is $f(x, y) = x + 6y$.

We've done the hard part. Now we plot $x \geq 0$, $y \geq 0$, $3x + 2y \geq 18$, and $x + 4y \geq 8$ on a graph and test the vertices.

(Small note. $x \geq 0$ since we cannot haul a negative amount of tub water.)

$3x + 2y = 18$ and $x + 4y = 8$ intersect at $(5.6, 0.6)$ I used Cramer's rule.

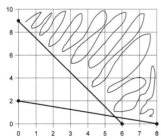

Testing the vertices in $f(x, y) = x + 6y$
$f(0, 9) = 54$
$f(5.6, 0.6) = 9.2$
$f(8, 0) = 8$ ⇦ the winner

To minimize my time I should use tub water.

The Complete Solutions and Answers

329. Prove $1 + 3 + 5 + 7 + \ldots + 2n - 1 = n^2$ for every natural number n.

The first step is to prove the n = 1 statement is true. This part of the math induction proof is usually super obvious.

$$n = 1 \Rightarrow 1 \stackrel{?}{=} 1^2 \quad \text{I told you it was obvious.}$$

The second step is to assume the n = k statement is true.
Namely, we assume $n = k \Rightarrow 1 + 3 + 5 + \ldots + 2k - 1 = k^2$.
We are allowed to use this fact in trying to prove the n = k+1 statement.

$$\text{To prove: } n = k + 1 \Rightarrow 1 + 3 + 5 + \ldots + 2(k+1) - 1 \stackrel{?}{=} (k+1)^2$$

Let's start with what we are assuming to be true:
$$1 + 3 + 5 + \ldots + 2k - 1 = k^2$$

What's the next number after $2k - 1$ in the series? These are odd numbers that are separated from each other by 2. If I add 2 to any number in the series, I get the next number.

The next number after $2k - 1$ is $2k - 1 + 2$.

$2k - 1 + 2$ simplifies to $2k + 1$.

If I add $2k + 1$ to both sides of the series I'm assuming to be true, I get:

$$1 + 3 + 5 + \ldots + 2k - 1 + \mathbf{2k + 1} = k^2 + \mathbf{2k + 1}$$

But this is exactly what I'm trying to show is true.

The details: $2k + 1$ is the same as $2(k + 1) - 1$

$k^2 + 2k + 1$ factors into $(k + 1)^2$

Done!

---

330. Most dogs like to fetch balls. Wufwuf is different. She likes to fetch pumpkins. One Halloween she ran around the neighborhood and stole 3 pumpkins and brought them home. She didn't tell me about her theft but hid the pumpkins behind the backyard wall.

On the second Halloween, she stole 12 pumpkins and added them to her stash behind the backyard wall.

On the third Halloween, she stole 48 pumpkins. Each year she got better at finding pumpkins and taking them. Each year she took 4 times as many as the previous year.

How many pumpkins were hidden behind the backyard wall after 8 Halloweens?

$a = 3, r = 4, n = 8$

$$s = \frac{a(1 - r^n)}{1 - r} = \frac{3(1 - 4^8)}{1 - 4} = \frac{3(1 - 65536)}{-3} = 65{,}535 \text{ pumpkins.}$$

The Complete Solutions and Answers

331. Graph by point-plotting y = x! from x = 0 to x = 5.

If x = 0, then y = 1. (Details: y = 0! = 1 using a calculator.) The point (0, 1).
If x = 1, then y = 1. We have the point (1, 1).
If x = 2, then y = 2. We have the point (2, 2).
Similarly we get the points (3, 6), (4, 24), (5, 120).

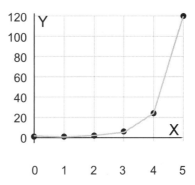

The factorial function climbs very fast. 10! is over three million. 40! has 48 digits. 60! is larger than the number of atoms in the observable universe.

You Want a Function that Really Climbs?

Imagine a bug on an infinitely long perforated paper tape—like a roll of postage stamps.

At the start, each square on the tape is white.

What the bug does depends on two things: ① What color the square he is on and ② What mood he is in.

① The squares can be either white or black.
② The bug's mood can either curious, glad, afraid, friendly, itchy, silly. . . .
(a finite list)
At any moment in time he is in exactly one of those states.

(continued next page)

The Complete Solutions and Answers

He always starts "the game" in the state of curious.

    Here is how the game is played. The bug looks at the color of the square he is on and looks at his mood. On the basis of those two things, he does three things: ❶ He colors the square he is on either white or black, ❷ He moves one square to the left or to the right, and ❸ He chooses a mood—curious, glad, afraid, friendly, itchy, silly . . . —or he decides to stop.

    For example, if the square he is on is black and his mood was curious, he might decide to color the square black, move one square to the left, and change his mood to glad.

    (Black, Curious) → (Black, Left, Glad)

    If he were on a black square and glad, he might decide to color the square black, move one square to the left, and change his mood to curious.

    (Black, Glad) → (Black, Left, Curious)

	Curious	Glad
White		
Black	color it Black, move Left, become Glad	color it Black, move Left, become Curious

Your role in the game is *to decide what the bug will do in each case.*

    If the bug had two moods (curious and glad), you might decide:

	Curious	Glad
White	color it Black, move Right, become Glad	color it White move Right Stop
Black	color it Black, move Left, become Glad	color it Black, move Left, become Curious

    Put the bug on the tape and watch him go. He would be on a white square to start with and he always starts in a curious mood.

    His first step would be to color the square Black, move Right, and become Glad. We know this by looking at the chart.

(continued next page)

The Complete Solutions and Answers

His second step (since he is on a white square and is glad) would be to color the square White, move Right, and stop.

That game would have lasted two steps.

Here is a game that would last forever and never stop.

	Curious	Glad
White	B, L, G	B, L, C
Black	B, R, C	B, L, Stop

His first step would be to color the square Black, move Left, and become Glad.

His second step (since he is on a white square and is glad) would be to color the square Black, move Left, and become Curious.

Third step (since he is on a white square and is curious) would be to color the square Black, move Left, and become glad.

And this would go on and on without ever hitting Stop.

The goal of the game is to design the chart so that the bug would make as many steps as possible and would at some point stop.

<u>For a bug with one mood</u>, the bug could make at most one step.

	Curious
White	B, Left, Stop
Black	B, Right, Curious

<u>For a bug with two moods</u>, using the best chart possible, the bug could make 6 steps before stopping.

	Curious	Glad
White	B, R, G	B, L, C
Black	B, L, G	B, R, Stop

You don't need to believe me. Just take out a piece of paper and draw the perforated tape roll ⋯ | | | | | | | ⋯ and watch the bug move 6 times before stopping.

(continued next page)

The Complete Solutions and Answers

There will be four black squares in a row when he stops.
☐☐☐☐☐☐☐■■■■☐☐☐☐☐☐☐☐☐

<u>For a bug with three moods</u>, using the best chart possible, the bug could make 14 steps before stopping.

When I studied this at the university, we were given the homework assignment of finding a chart that would make this happen. It took me over an hour to create the chart.

So far we have:
- 1 mood → 1 step
- 2 moods → 6 steps
- 3 moods → 14 steps

If you were graphing this by point-plotting, it would look like:

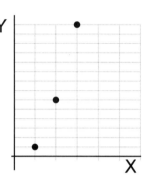

<u>For a bug with four moods</u>, using the best chart possible, the bug could make 107 steps before stopping.

That function is climbing!

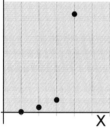

<u>For a bug with five moods</u>, we have NO IDEA what the best chart is. No one has figured it out yet. Heiner Marxen and Jürgen Buntrock found a chart back in 1989 that made the bug move 47,176,870 steps before stopping. So we know that for five moods, there are at least 47,176,870 steps.

<u>For a bug with six moods</u>, Pavel Kropitz in 2010 found a chart that would make the bug move over 10^{36534} steps. It is almost certainly not the best possible chart. 2010 is not that long ago. This is very recent mathematics.

(continued next page)

The Complete Solutions and Answers

So for a bug with six moods (curious, glad, afraid, friendly, itchy, and silly) we now have a chart with 12 boxes

	Curious	Glad	Afraid	Friendly	Itchy	Silly
White						
Black						

that would make the bug move over 1,000,000,000,000,000,000,000,000,000,000,000,000,000,000,000,000,000,000,000,000,000,000,000,000,000,000,000,000,000,000,000,000,000,000,000,000,000,000,000,000,000,000,000,000...

The Complete Solutions and Answers

000,000,000,000,000,000,000,000,000,000,000,000,000,000,000,000,000,000,000,000,000,000,000,000,000,000,000,000,000,000,000,000,000,000,000,000,000,000,000,000,000,000,000,000,000,000,000,000,000,000,000,000,000,000,000,000,...

[a large block of "000,000,..." digits representing approximately 10^{36534}]

...0,000

steps before stopping. (Yes. There are 36,534 zeros listed in the above number. I counted them.)

$1 \to 1$
$2 \to 6$
$3 \to 14$
$4 \to 107$
$5 \to$ at least $47{,}176{,}870$
$6 \to$ at least 10^{36534}

is what could be considered a function that really climbs.

(In the math world, this function is called the busy beaver function.)

The Complete Solutions and Answers

333. Evaluate $\begin{vmatrix} 3 & 5 \\ 7 & 4 \end{vmatrix} = (3)(4) - (7)(5) = -23$

335. Solve for z. You do not have to evaluate the determinants.

$$\begin{cases} 2x + y + 8z = 9 \\ 7x + 3y - 5z = 8 \\ 3x - 8y + z = 7 \end{cases} \qquad z = \frac{D_z}{D}$$

$$z = \frac{\begin{vmatrix} 2 & 1 & 9 \\ 7 & 3 & 8 \\ 3 & -8 & 7 \end{vmatrix}}{\begin{vmatrix} 2 & 1 & 8 \\ 7 & 3 & -5 \\ 3 & -8 & 1 \end{vmatrix}}$$

337. Is f a function?

$$\text{cuff} \xrightarrow{f} \text{❋}$$
$$\text{sleeve} \xrightarrow{f} \text{rock}$$
$$\text{shoe} \xrightarrow{f} \text{❋}$$

Does cuff have exactly one image? Yes.
Does sleeve have exactly one image? Yes.
Does shoe have exactly one image? Yes.

    Yes. It is a function. In determining whether something is a function, we have to look at each element in the domain.

    The fact that both cuff and shoe have the same image has nothing to do with whether f is a function.

    Another way to name a function is to write   f(cuff) = ❋
                                                                      f(sleeve) = rock
                                                                       f(shoe) = ❋

    In English this is read as, "f of cuff is equal to ❋."

    It is the function called f operating on the element cuff in the domain and yielding its image ❋ in the codomain.

    This is not f *times* cuff. It is really tough to confuse f(cuff) with 5(x + 3y).

| 339–345 | The Complete Solutions and Answers |

339. Given points (x_1, y_1) and (x_2, y_2) on line ℓ. Find the length of the segment that joins those two points.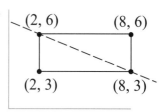
❶ Find the coordinates of C.
   (x_2, y_1) B and C have the same abscissas.
              A and C have the same ordinates.
❷ Find the length of BC.
   $y_2 - y_1$ From (x_2, y_1) to (x_2, y_2)
❸ Find the length of AC.
   $x_2 - x_1$ From (x_1, y_1) to (x_2, y_1)
❹ Find the length d of the segment joining (x_1, y_1) and (x_2, y_2) by using the Pythagorean theorem.
$$(x_2 - x_1)^2 + (y_2 - y_1)^2 = d^2$$
$$\sqrt{(x_2 - x_1)^2 + (y_2 - y_1)^2} = d$$

> The distance between two points (x_1, y_1) and (x_2, y_2)
> $$d = \sqrt{(x_2 - x_1)^2 + (y_2 - y_1)^2}$$

341. $\dfrac{\sqrt{x} + 5}{\sqrt{x}} = \dfrac{\sqrt{x} + 5}{\sqrt{x}} \cdot \dfrac{\sqrt{x}}{\sqrt{x}} = \dfrac{(\sqrt{x} + 5)\sqrt{x}}{x}$ OR $\dfrac{x + 5\sqrt{x}}{x}$

343. First, fill in the coordinates of the other two vertices of the rectangle.
Then use the two-point form of the line.
$$\frac{y - 6}{x - 2} = \frac{3 - 6}{8 - 2}$$
$$\frac{y - 6}{x - 2} = \frac{-3}{6} \quad \text{or} \quad x + 2y = 14$$

(2, 6)   (8, 6)
(2, 3)   (8, 3)

345. $\dfrac{\log_3 7}{\log_3 2} = \log_2 7$

$(\log_\pi 8)(\log_8 23) = \log_\pi 23$

One way that I remember $(\log_c b)(\log_b a) = \log_c a$ is that I imagine somehow canceling the b $\log_b$: $(\log_c \cancel{b})(\cancel{\log_b} a) = \log_c a$. It's just a memory aid. It's not real canceling.

124

The Complete Solutions and Answers 347–350

347. $\dfrac{(x-7)^2}{81} + \dfrac{(y+2)^2}{100} = 1$ is an ellipse centered at (7, –2) with a semi-major axis length of 10 and a semi-minor axis length of 9. How would $\dfrac{(x-7)^2}{100} + \dfrac{(y+2)^2}{100} = 1$ be described?

The first thing that makes this "ellipse" unusual is that its semi-major axis and semi-minor axis are equal. That makes a very round ellipse. In other words, it is a circle.

With equations for circles, you don't need to have fractions. Instead of $\dfrac{(x-7)^2}{100} + \dfrac{(y+2)^2}{100} = 1$ we write $(x-7)^2 + (y+2)^2 = 10^2$

This is a circle with center at (7, –2) and radius equal to 10.

349. Which of these are graphs of functions?

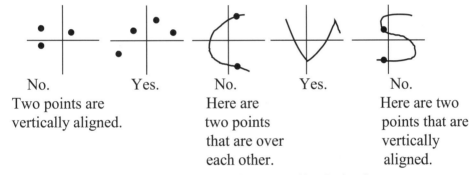

No. Yes. No. Yes. No.
Two points are vertically aligned. Here are two points that are over each other. Here are two points that are vertically aligned.

From the artistic point of view, if you can draw a vertical line that hits the curve in two or more different points, then that is the graph of something that is not a function.

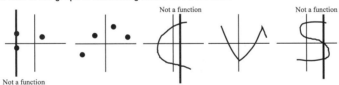

350. Express in sigma notation $\log(3x^2) + \log(3x^3) + \log(3x^4) + \log(3x^5)$

There are (at least) two possibilities:

$$\sum_{i=1}^{4} \log(3x^{i+1}) \quad \text{or} \quad \sum_{i=2}^{5} \log(3x^i)$$

351–365 The Complete Solutions and Answers

351. The time (t) it takes to plow a square field varies directly as the square of the length of one of the sides (s).

$$t = ks^2$$

353. $\log_{36} 6 = \frac{1}{2}$ since $36^{1/2} = \sqrt{36} = 6$
$\log_{\sqrt{7}} 7 = 2$ since $(\sqrt{7})^2 = 7$
$\log_{1000} 10 = \frac{1}{3}$ since $1000^{1/3} = \sqrt[3]{1000} = 10$

359. Graph $\dfrac{(x-3)^2}{49} - \dfrac{(y-4)^2}{1} = 1$

Draw the egg box. X it out. Draw the hyperbola. Since the x² term is positive, the curve opens to the left and right (rather than vertically).

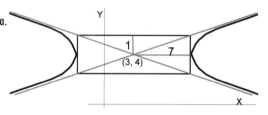

365. Let f:N → N be defined by f(x) = x + 7 where N is the set of natural numbers {1, 2, 3, 4, 5, 6 . . . }. Show that f^{-1}: N → N is *not* a function.

If f^{-1} were a function, then, by definition, every element in the domain of f^{-1} must have exactly one image in the domain. This works fine for $f^{-1}(16)$. It is equal to 9, since f(9) = 16.

However, $f^{-1}(5)$ has no image in N. There is no number in N such that adding 7 to it will give an answer of 5.

In this case, f is a function, but f^{-1} is not a function.

Here is an example in a picture.

g is a function.

$g^{-1}(\mathbb{N})$ does not exist.

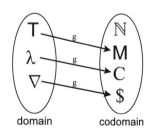

The Complete Solutions and Answers

370. Prove $1^3 + 2^3 + 3^3 + 4^3 + \ldots + n^3 = \dfrac{n^2(n+1)^2}{4}$ is true for every natural number.

The first step is to prove the n = 1 statement is true.

$$n = 1 \Rightarrow 1^3 \stackrel{?}{=} \dfrac{1^2(1+1)^2}{4} \quad \text{True (and obvious)}$$

The second step is to assume the n = k statement is true.
Namely, we assume $n = k \Rightarrow 1^3 + 2^3 + 3^3 + 4^3 + \ldots + k^3 = \dfrac{k^2(k+1)^2}{4}$

We are allowed to use this fact in trying to prove the n = k+1 statement.

To prove: $n = k+1 \Rightarrow 1^3 + 2^3 + \ldots + (k+1)^3 = \dfrac{(k+1)^2(k+2)^2}{4}$

Let's start with what we are assuming to be true:

$$1^3 + 2^3 + 3^3 + 4^3 + \ldots + k^3 = \dfrac{k^2(k+1)^2}{4}$$

Add $(k+1)^3$ to each side:

$$1^3 + 2^3 + 3^3 + 4^3 + \ldots + k^3 + (k+1)^3 = \dfrac{k^2(k+1)^2}{4} + (k+1)^3$$

The left side of this equation is the left side of what we are trying to prove.

I have to turn $\dfrac{k^2(k+1)^2}{4} + (k+1)^3$ into $\dfrac{(k+1)^2(k+2)^2}{4}$ and I'll be done.

$$\dfrac{k^2(k+1)^2}{4} + (k+1)^3$$

$$= \dfrac{k^2(k+1)^2}{4} + \dfrac{4(k+1)^3}{4} \quad \text{Adding fractions like we did back in Chapter 4½}$$

$$= \dfrac{k^2(k+1)^2 + 4(k+1)^3}{4}$$

$$= \dfrac{(k+1)^2(k^2 + 4(k+1))}{4} \quad \text{Factoring } (k+1)^2 \text{ from each term. It is a little like what we did in factoring by grouping. The alternative would be to multiply everything out, but I don't want to do all that work.}$$

$$= \dfrac{(k+1)^2(k^2 + 4k + 4)}{4}$$

$$= \dfrac{(k+1)^2(k+2)^2}{4} \quad k^2 + 4k + 4 \text{ is a perfect square. See page 428 in Life of Fred: Beginning Algebra Expanded Edition.}$$

Done!

| 371–383 | The Complete Solutions and Answers

371. When pirates attack a ship, the loot (L) each of them gets varies inversely as the number of pirates (p) doing the stealing. If 16 pirates attack the *Pinta*, they would each get 60 pieces of silver. If 12 pirates attack the *Pinta*, how many pieces of silver would each of them get?

Step ①: Find the equation. $L = \dfrac{k}{p}$

Step ②: Find the value of k.   16 pirates get 60 pieces of silver

Substituting that into $L = \dfrac{k}{p}$ we get $60 = \dfrac{k}{16}$

Multiply both sides by 16 $960 = k$

$L = \dfrac{k}{p}$ now becomes $L = \dfrac{960}{p}$

Step ③: Find L when p = 12.

$$L = \dfrac{960}{12}$$

$$L = 80$$

The twelve pirates would each receive 80 pieces of silver and would have probably messed up Columbus's voyage.

377. Draw a Venn diagram of the set of all cats (C) and the set of all pizzas (P).

Since no cat is a pizza and no pizza is a cat, the sets are have no element in common. $C \cap P = \varnothing$

When two sets have no element in common, we say that the sets are **disjoint**.

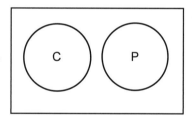

383. At what point does $\dfrac{x}{5} + \dfrac{y}{6} = 1$ intercept the x-axis?

Answer: (5, 0)

Note that if you put x = 5 into the equation, you get $1 + \dfrac{y}{6} = 1$, so y would have to be zero.

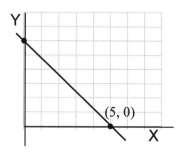

The Complete Solutions and Answers

389. Factor $x^3 + 7x^2 + 3x + 21$
$$= x^2(x + 7) + 3(x + 7)$$
$$= (x + 7)(x^2 + 3)$$

390. $\dfrac{-2x - 26}{x^2 - 4x - 21}$

To make this look like the previous problem, factor the denominator.

    One of the big differences between studying mathematics and studying art history is that we actually use the material that we learned earlier in the course. In this problem, we will have to factor the "easy trinomial" $x^2 - 4x - 21$. We want two numbers that add to -4 and multiply to -21. $(x + 3)(x - 7)$.

    In art history, if you forget everything you learned about painters in the 15th century, you can still study 18th century painters.

    In contrast, in math it is actually important to remember the earlier stuff. You need to know your addition and multiplication tables in order to factor trinomials. You need to know how to factor trinomials in order to do partial fractions. You will need to know how to do partial fractions in calculus. At my local university, you need to prove that you know calculus to enter the masters in business administration (MBA) program.

    Parenthetically, I have taken courses in art history and have enjoyed them—except for having to memorize the dates of the works of art. Learning the multiplication tables involved a lot fewer items to memorize, and they were much more useful.

$$\frac{-2x - 26}{x^2 - 4x - 21} = \frac{-2x - 26}{(x + 3)(x - 7)} = \frac{A}{x + 3} + \frac{B}{x - 7}$$

Multiply each term by $(x + 3)(x - 7)$

$$\frac{(-2x - 26)(x + 3)(x - 7)}{(x + 3)(x - 7)} = \frac{A(x + 3)(x - 7)}{x + 3} + \frac{B(x + 3)(x - 7)}{x - 7}$$

All the denominators disappear.

$$-2x - 26 = A(x - 7) + B(x + 3)$$

Let $x = 7$ $-2(7) - 26 = B(10)$
$$-4 = B$$

Let $x = -3$ $-2(-3) - 26 = A(-3 - 7)$
$$2 = A$$

$$\frac{-2x - 26}{x^2 - 4x - 21} = \frac{2}{x + 3} + \frac{-4}{x - 7}$$

| 395 | The Complete Solutions and Answers

395. Graph $y > \frac{3}{4}x + 2$

First, graph the equality $y = \frac{3}{4}x + 2$

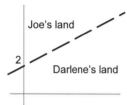

Using $y = mx + b$

If the original problem is $\geq$ or $\leq$, use a solid line.
If the original problem is $>$ or $<$, use a dashed line.
Second, pretend that the line you have just graphed is a fence that separates pastures owned by different people.

You need to test one bite of grass in each owner's pasture to see where the grass is good. Translation: You need to test one point in Joe's land and one point in Darlene's land to see if it satisfies the original inequality.

I, your reader, have a question. How do I find a point in Joe's land?

You just have to look at the graph you have drawn. How about the point (0, 5)?

And for Darlene's land?

How about (0, 0). That certainly is on her side of the fence. Or you could choose (8, 0) or (39723, 1) or (0, −45). They are all obviously on Darlene's land. I choose (0, 0) because I like easy numbers. There is less arithmetic.

Testing (0, 5), which is on Joe's land:
Put (0, 5) into the original inequality. $5 \stackrel{?}{>} \frac{3}{4}(0) + 2$

This is true. Therefore, every point on Joe's land satisfies the original inequality.

Testing (0, 0), which is on Darlene's land:
Put (0, 0) into the original inequality. $0 \stackrel{?}{>} \frac{3}{4}(0) + 2$

This is false. So no point on Darlene's land satisfies the original inequality.

Third, we shade in the regions that tasted good.

Usually one region is shaded in. Sometimes both are shaded. Sometimes, neither.

The Complete Solutions and Answers | 397

397. When I first moved into my house years ago, I knew that I had to store some of my math books and shoes in the garage.

The red moving box could hold 12 math books and 6 shoes. The blue moving box could hold 12 math books and 10 shoes.

I knew that I needed to store at least 96 math books and 60 shoes. Should I use red boxes, blue boxes, or some of each? Red boxes take up 9 cubic feet. Blue boxes, 10. I want to minimize the volume.

We want to minimize the volume, but the minimum volume is *not* what we are looking for. We want to know how many red boxes and how many blue boxes I should use.

Let x = number of red boxes I'll use.

Let y = number of blue boxes I'll use. The start of every linear algebra problem should have two "Let x or y equal" statements.

Reading the English and the previous three lines . . .

$12x + 12y \geq 96$ for the math books

$6x + 10y \geq 60$ for the shoes

Since red boxes take up 9 cubic feet and blue boxes, 10, the total volume is $9x + 10y$, which I want to minimize.

The objective function is $f(x, y) = 9x + 10y$.

Plot $x \geq 0$, $y \geq 0$, $12x + 12y \geq 96$, and $6x + 10y \geq 60$ on a single graph and test the vertices.

$12x + 12y = 96$ and $6x + 10y = 60$ intersect at $(5, 3)$. I used the elimination method.

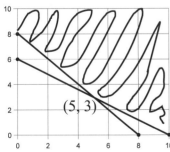

$f(x, y) = 9x + 10y$

Testing the vertices . . .

$f(0, 8) = 80$

$f(5, 3) = 75$ ⇐ the winner

$f(10, 0) = 90$

I should use 5 red boxes and 3 blue boxes. They will take up 75 cubic feet. That will leave room so that I can park my car in the garage.

| 401–407 | **The Complete Solutions and Answers**

401. $(8^x)^y = 8^{xy}$
$9^{-2} = 1/9^2 = 1/81$
$1/a^{-7} = a^7$

407. $\log_5 \sqrt{5} = \frac{1}{2}$ since 5 raised to the one-half power is $\sqrt{5}$

In Chapter 1½ we worked with fractional exponents. $x^{1/n} = \sqrt[n]{x}$

$\log_3 \frac{1}{3} = -1$ since $3^{-1} = \frac{1}{3}$

In Chapter 1½ we worked with negative exponents. $x^{-a} = \frac{1}{x^a}$

$\log_{82} 1 = 0$ since any number to the zero power equals 1.

In Chapter 1½ we worked with zero exponents. $x^0 = 1$

    Chapter 1½ was labeled "Looking Back" in *Life of Fred: Advanced Algebra* because it was reviewing the material taught in Beginning Algebra, but it could have been labeled "Looking Forward" because we were going to need it in Chapter 3 on logarithms.

    One of the hard things about teaching is knowing where the students will use the material, but not being able to tell them at the time it's being taught. Teaching logs is a good example. When we teach it in this chapter, we show that it is good for solving exponential equations like $2^{x+3} = 7.8$, but one of the major uses of logs will occur in second-semester calculus.

    In that calculus course we prove that:

the anti-derivative of x^7 is $(1/8)x^8$ (Whatever anti-derivative means!)
the anti-derivative of x^6 is $(1/7)x^7$
the anti-derivative of x^5 is $(1/6)x^6$
the anti-derivative of x^4 is $(1/5)x^5$
the anti-derivative of x^3 is $(1/4)x^4$
the anti-derivative of x^2 is $(1/3)x^3$
the anti-derivative of x^1 is $(1/2)x^2$
the anti-derivative of x^0 is $(1/1)x^1$
the anti-derivative of x^{-1} is $\log_{2.71828182845904523536028747713527} x$
the anti-derivative of x^{-2} is $(1/(-1))x^{-1}$

    Did you notice that one of the anti-derivatives is a bit funny? Not only is it a log, but it has a very weird-looking base equal to 2.71828182845904523536028747713527. The anti-derivative of $\frac{1}{x}$ is always equal to $\log_{2.71828182845904523536028747713527} x$ and not $\log_{2.74} x$ or $\log_{2.59} x$.

    This is one major use of logs, but I can't mention it right now. ☺

The Complete Solutions and Answers 413–425

413. Solve $3^{x+2} = 7$

$\log 3^{x+2} = \log 7$ — Take the log of both sides

When no base is indicated, log means $\log_{10}$. $\log_{10}$ are called common logs.

$(x+2)\log 3 = \log 7$ — The Birdie Rule

$x + 2 = \dfrac{\log 7}{\log 3}$ — Divide both sides by log 3

$x = \dfrac{\log 7}{\log 3} - 2$ — Subtract 2 from both sides

419. Solve $\begin{cases} 4x + 3y = 8 \\ 6x + y = 5 \end{cases}$

This could be done by either the elimination method or by the substitution method. Since there is a plain y in the second equation, I will do it by substitution.

Solve the second equation for y $y = 5 - 6x$

Substitute $y = 5 - 6x$ into the first equation $4x + 3(5 - 6x) = 8$

$$4x + 15 - 18x = 8$$
$$7 = 14x$$
$$\tfrac{1}{2} = x$$

Back substitute $x = \tfrac{1}{2}$ into $y = 5 - 6x$ $y = 5 - 6(\tfrac{1}{2})$

$$y = 2$$

425. Put in the standard form for a circle $(x - h)^2 + (y - k)^2 = r^2$ and graph:

$16x^2 + 32x + 16y^2 - 96y = 9$

$16(x^2 + 2x) + 16(y^2 - 6y) = 9$ x^2 and y^2 must have a coefficient of 1

$16(x^2 + 2x + 1) + 16(y^2 - 6y + 9) = 9 + 16 + 144$

$16(x + 1)^2 + 16(y - 3)^2 = 169$

$(x + 1)^2 + (y - 3)^2 = \dfrac{169}{16}$

$(x - (-1))^2 + (y - 3)^2 = \left(\dfrac{13}{4}\right)^2$

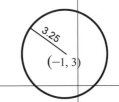

A circle centered at $(-1, 3)$ with radius equal to $\dfrac{13}{4}$

The Complete Solutions and Answers

431. Solve $\dfrac{x+1}{7} = \dfrac{5x-3}{21}$

Cross multiplying	$21(x+1) = 7(5x-3)$
Distributive law	$21x + 21 = 35x - 21$
Subtract 21x from both sides	$21 = 14x - 21$
Add 21 to both sides	$42 = 14x$
Divide both sides by 14	$3 = x$

437. Draw a Venn diagram of the intersection of the set of all weasels in the world (W) and all animals in America (A).

Since there are weasels that are in America, the sets are not disjoint. $W \cap A \neq \emptyset$.

Since there are weasels that are not in America, $W \subset A$ is not true.

Since there are animals in America that are not weasels, $A \subset W$ is also not the case.

W ∩ A

443. Find the center of the ellipse whose vertices are (5, 8) and (5, 12). Find the length of the semi-major axis. Can you find the length of the semi-minor axis?

The center is halfway between the two vertices. Halfway between 8 and 12 is the average of 8 and 12, which is 10. (8 plus 12 and divide by 2). The center is at (5, 10).

The semi-major axis is the distance between the center and either of the vertices. The distance from (5, 10) to (5, 12) is 2.

Not enough information is given to determine the length of the semi-minor axis. The ellipse could be fat or skinny.

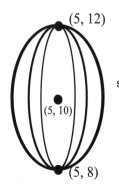

Here are three ellipses with the same pair of vertices.

The Complete Solutions and Answers

445. Is this a function? Is it 1-1? Is it onto?

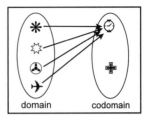

It is a function since each member of the domain has exactly one image in the codomain.

It is not 1-1 since ☺ is the image of more than one element of the domain.

It is not onto since ✚ is not the image of any element of the domain.

446. Let the domain and codomain of function f be the integers {. . . –3, –2, –1, 0, 1, 2, 3, 4 . . .}. Define f by $f(x) = x^2$.

Is f a 1-1 function?

$f(–5) = 25$ and $f(5) = 25$. This shows that f is not 1-1. There are a million examples possible. All I had to show was one example in order to prove that f isn't 1-1.

450. Unadd $\dfrac{46x - 3}{15x^2 - x - 2}$

First, factor the denominator. It's a general trinomial. We want two things that add to $-x$ and multiply to $-30x^2$ ($15x^2$ times -2).

$$15x^2 - x - 2$$
$$= 15x^2 - 6x + 5x - 2$$

factor now by grouping
$$= 3x(5x - 2) + 1(5x - 2)$$
$$= (5x - 2)(3x + 1)$$

$$\dfrac{46x - 3}{15x^2 - x - 2} = \dfrac{46x - 3}{(5x - 2)(3x + 1)} = \dfrac{A}{5x - 2} + \dfrac{B}{3x + 1}$$

Multiply each term by $(5x - 2)(3x + 1)$ to eliminate the denominators

$$46x - 3 = A(3x + 1) + B(5x - 2)$$

What do I let x equal in order to eliminate the B term? I want $5x - 2$ to equal zero. $5x - 2 = 0$
                                                                                          $5x = 2$
Let $x = \dfrac{2}{5}$ $46(\dfrac{2}{5}) - 3 = A(3(\dfrac{2}{5}) + 1)$ $x = 2/5$

$$15.4 = 2.2A$$
$$7 = A$$

I can work in decimals or in fractions. It's my choice.

Let $x = -\dfrac{1}{3}$ $46(-\dfrac{1}{3}) - 3 = B(5(-\dfrac{1}{3}) - 2)$

$$\dfrac{-55}{3} = \dfrac{-11}{3} B$$
$$5 = B$$

$$\dfrac{46x - 3}{(5x - 2)(3x + 1)} = \dfrac{7}{5x - 2} + \dfrac{5}{3x + 1}$$

455–467 The Complete Solutions and Answers

455. Simplify $\sqrt{50}$ $\sqrt{50} = \sqrt{25}\sqrt{2} = 5\sqrt{2}$

461. Graph $2x + 4y = 8$

If $x = 0$, then $y = 2$. (details: $2(0) + 4y = 8$) We have the point $(0, 2)$.
If $x = 2$, then $y = 1$. We have the point $(2, 1)$.
If $x = 4$, then $y = 0$. We have the point $(4, 0)$.

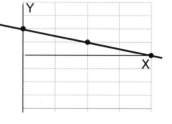

467. Solve $\sqrt{22 - 3x} + 3 = 7$

First, isolate the square root. $\sqrt{22 - 3x} = 4$
Second, square both sides. $22 - 3x = 16$
Solve. $6 = 3x$
                                    $2 = x$

Third, check the answer in the original equation.
Checking $x = 2$
$$\sqrt{22 - 6} + 3 \stackrel{?}{=} 7$$
$$\sqrt{16} + 3 \stackrel{?}{=} 7 \quad \text{yes}$$

The solution to $\sqrt{22 - 3x} + 3 = 7$ is $x = 2$.

470. Graph $y = x^2 - 4x + 12$ from $x = 1$ to $x = 3$. Please plot at least five points.

I will be silly and point-plot a zillion points.

x	y
1	9
1.2	8.64
1.4	8.36
1.6	8.16
1.8	8.04
2	8
2.2	8.04
2.4	8.16
2.6	8.36
2.8	8.64
3	9

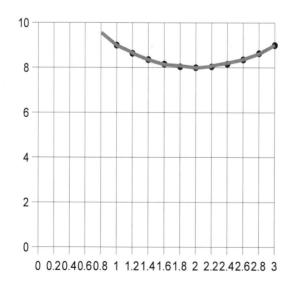

The Complete Solutions and Answers

473. Graph $y > 3x^2 + 2x$

Draw the fence. (Graph the equality.)

Point-plotting $y = 3x^2 + 2x$

$x = 0, y = 0$
$x = 1, y = 5$
$x = -1, y = 1$
$x = 2, y = 16$
$x = -2, y = 8$

Dashed line since the original inequality was $>$ and not $\geq$.

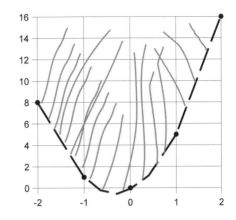

Test a point in each region.

$(0, 12)$ is above the curve. $12 \overset{?}{>} 3(0)^2 + 2(0)$ is true.

$(0, -7)$ is below the curve. $-7 \overset{?}{>} 3(0)^2 + 2(0)$ is false.

Shade the area above the curve.

479. Is g a function?

g(rice) = can
g(mouse) = can
g(Harvard) = can
g(clock) = can
g(shovel) = rooster

The domain of this function is {rice, mouse, Harvard, clock, shovel}. Each element of the domain has exactly one image under function g. That is all we need to know in order to say that g is a function.

A function is *any* rule which associates to each element of the domain, exactly one image in the codomain. *Any* rule.

I could invent another function and call it h:

h(rice) = can
h(mouse) = rooster
h(Harvard) = can
h(clock) = rooster
h(shovel) = rooster

Functions g and h both have the same domains, but they are different functions. They are different rules that make different assignments.   g(mouse) = can   h(mouse) = rooster

The Complete Solutions and Answers

485. The life of a car (L) measured in months varies inversely as the average number of miles (m) that it is driven each day.

$$L = \frac{k}{m}$$

491. $\dfrac{4}{3+x+\sqrt{y}} = \dfrac{4}{3+x+\sqrt{y}} \cdot \dfrac{3+x-\sqrt{y}}{3+x-\sqrt{y}}$

$= \dfrac{4(3+x-\sqrt{y})}{9+6x+x^2-y}$

The details: $(3+x+\sqrt{y})(3+x-\sqrt{y})$ is in the form $(a+b)(a-b)$, which equals $a^2 - b^2$.

So $(3+x+\sqrt{y})(3+x-\sqrt{y})$ equals $(3+x)^2 - y$,

and $(3+x)^2 = (3+x)(3+x) = 9 + 3x + 3x + x^2 = 9 + 6x + x^2$

497. $i^5 = i^4 \cdot i = 1i = i$

$3 + 4i + 7 - 6i = 10 - 2i$

$(2+5i)(8+i) = 16 + 2i + 40i + 5i^2 = 16 + 42i - 5 = 11 + 42i$

500. Factor $\quad 3x^2y - xyw + 3xw - w^2$

$= xy(3x - w) + w(3x - w)$

$= (3x - w)(xy + w)$

$45x + 15xy + 6y + 2y^2$
$= 15x(3 + y) + 2y(3 + y)$
$= (3 + y)(15x + 2y)$

$6ac + 12ad + c + 2d$
$= 6a(c + 2d) + (c + 2d)$
$= (c + 2d)(6a + 1)$

$8xy + 20x + 6y + 15$
$= 4x(2y + 5) + 3(2y + 5)$
$= (2y + 5)(4x + 3)$

The Complete Solutions and Answers 503–509

503. Simplify as much as possible $(\log_{10} 2)(\log_2 100)$
$$= \log_{10} 100$$
$$= 2$$

509. Graph $y = \ln x$ from $x = 1$ to $x = 10$

If $x = 1$, then $y = 0$. (details: $y = \ln 1 = 0$)   We have the point $(1, 0)$.
If $x = 2$, then $y \approx 0.7$ (details: punch in 2 and hit the ln key) $(2, 0.7)$
If $x = 3$, then $y \approx 1.1$ $(3, 1.1)$
If $x = 5$, then $y \approx 1.6$ $(5, 1.6)$
If $x = 7$, then $y \approx 1.9$ $(7, 1.9)$
If $x = 10$, then $y \approx 2.3$ $(10, 2.3)$

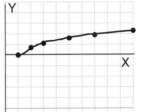

Now, we can divulge what the ln function is.

Remember that log x means $\log_{10} x$.

The function ln x is defined as $\log_e x$.

e is the irrational number which is approximately equal to 2.71828182845904523536028747 13527

The other irrational number that you know about is π which is approximately equal to 3.1415926535897932
3846264338327950288419716939937510582097494459230781640628620899862803482534211706798214808651328230664709384460955058223172535940812848111745
0284102701938521105559644622948954930381964428810975665933446128475648233786783165271201909145648566923460348610454326648213393607260249141273
7245870066063155881748815209209628292540917153436789259036001133053054882046665213841469519415116094330572703657595919530921861173819326117931
0511854807446237996274956735188575272489122793818301194912983367336244065664308602139494639522473719070217986094370277053921717629317675238467
4818467669405132000568127145263560827785771342757789609173637178721468440901224953430146549585371050792279689258923542019956112129021960864034
4181598136297747713099605187072113499999983729780499510597317328160963185950244594553469083026425223082533446850352619311881710100031378387528
8658753320838142061717766914730359825349042875546873115956286388235378759375195778185778053217122680661300192787661119590921642019893809525720
1065485863278865936153381827968230301952035301852968995773622599941389124972177528347913151557485724245415069595082953311686172785588907509838 1
75463746493931925506040092770167113900984882401285836160356370766010471018194295559619846767837449948255379774726847104047534646208046682590

roughly speaking. π is defined as the circumference of any circle divided by its diameter.

The definition of e is a little harder.

Consider $(1 + \frac{1}{n})^n$ for $n = 1, 2, 3, 4, 5, \ldots$

When $n = 1$, $(1 + \frac{1}{n})^n$ equals 2

When $n = 2$, $(1 + \frac{1}{n})^n$ equals 2.25

When $n = 3$, $(1 + \frac{1}{n})^n$ approximately equals 2.37037

When $n = 4$, $(1 + \frac{1}{n})^n$ approximately equals 2.441406

The constant e is defined as $(1 + \frac{1}{n})^n$ when n gets infinitely large.

We will prove in calculus that if you raise e to the $i\pi$ power you get a really surprising answer:
$$e^{i\pi} = -1 \qquad \text{where } i = \sqrt{-1}$$

| 510–517 | The Complete Solutions and Answers |

510. What is the range of f?
 f(L) = M
 f(ξ) = π
 f(ℵ) = π

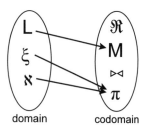

The set of images is {M, π}.
{M, π} is the range of f.

515. Addition is a function. The domain is the set of all ordered pairs of numbers and the codomain is the set of numbers. Is addition a 1-1 function?

 It is not a 1-1 function.
 Here is one counterexample: (2, 5) $\xrightarrow{+}$ 7
                              (3, 4) $\xrightarrow{+}$ 7

 All that is needed is one counterexample in order to show that a function if not 1-1.

 It is much harder to prove that a function *is* 1-1. You need to show that no two different elements in the domain have the same image.

Suppose you had a function that assigns to each person their full name. If the domain of this function were the people who live in Lost Springs, Wyoming, it would be pretty easy to verify that this function is 1-1. The 2010 U.S. census showed that the city population was 4. No two people in that city have the same full name. No two people are mapped by the function to the same name. That would be easy to check.

In contrast, if the domain of this function that assigns to each person their full name were the city of San Francisco, it is probable that the function would not be 1-1. Finding two people with the same full name would show that the function was not 1-1.

517. Describe the largest possible domain for the function whose rule is f(x) = √x .

 For most advanced algebra students, the correct answer will be that the largest possible domain for f(x) = √x will be the set of all non-negative numbers. (Not the set of all positive numbers, because √0 is equal to zero. √0 makes sense.)

Once we get to trig, we will find the square roots of negative numbers. You may have already seen √−1 = i. At that point, the largest possible domain would be the set of all real numbers.

However, we are not done. By the end of trig, you will find √i and the largest possible domain for f(x) = √x will be the set of all numbers—real, imaginary (6i), or complex (38 + 5i).

Have you ever wondered what √3 + 4i is equal to? Stay tuned.

The Complete Solutions and Answers | 521

521. $6x^5 + 7x - 12x^4 - 14 \div x - 2$

Arrange the terms in descending degree
Insert missing terms

$$x - 2 \overline{\smash{)}\, 6x^5 - 12x^4 + 0x^3 + 0x^2 + 7x - 14}$$

x into $6x^5$

$$\begin{array}{r} 6x^4 \\ x - 2 \overline{\smash{)}\, 6x^5 - 12x^4 + 0x^3 + 0x^2 + 7x - 14} \end{array}$$

$6x^4$ times $x - 2$

$$\begin{array}{r} 6x^4 \\ x - 2 \overline{\smash{)}\, 6x^5 - 12x^4 + 0x^3 + 0x^2 + 7x - 14} \\ 6x^5 - 12x^4 \end{array}$$

Subtract and bring
down the $0x^3$

$$\begin{array}{r} 6x^4 \\ x - 2 \overline{\smash{)}\, 6x^5 - 12x^4 + 0x^3 + 0x^2 + 7x - 14} \\ \underline{6x^5 - 12x^4} \\ 0x^4 + 0x^3 \end{array}$$

x into $0x^4$
$0x^3$ times $x - 2$

$$\begin{array}{r} 6x^4 \;\; + 0x^3 \\ x - 2 \overline{\smash{)}\, 6x^5 - 12x^4 + 0x^3 + 0x^2 + 7x - 14} \\ \underline{6x^5 - 12x^4} \\ 0x^4 + 0x^3 \\ 0x^4 + 0x^3 \end{array}$$

Subtract
Bring down $0x^2$
Divide x into $0x^2$

$$\begin{array}{r} 6x^4 \;\; + 0x^3 \\ x - 2 \overline{\smash{)}\, 6x^5 - 12x^4 + 0x^3 + 0x^2 + 7x - 14} \\ \underline{6x^5 - 12x^4} \\ 0x^4 + 0x^3 \\ \underline{0x^4 + 0x^3} \\ 0x^3 \;\; + 0x^2 \end{array}$$

x into $0x^3$
$0x^2$ times $x - 2$
Subtract and bring
down $+7x$

$$\begin{array}{r} 6x^4 \;\; + 0x^3 \;\; + 0x^2 \\ x - 2 \overline{\smash{)}\, 6x^5 - 12x^4 + 0x^3 + 0x^2 + 7x - 14} \\ \underline{6x^5 - 12x^4} \\ 0x^4 + 0x^3 \\ \underline{0x^4 + 0x^3} \\ 0x^3 + 0x^2 \\ \underline{0x^3 + 0x^2} \\ 0x^2 + 7x \end{array}$$

141

The Complete Solutions and Answers

x into $0x^2$

$0x$ times $x-2$

Subtract and bring down -14

$$\begin{array}{r} 6x^4 + 0x^3 + 0x^2 + 0x \\ x-2 \overline{) 6x^5 - 12x^4 + 0x^3 + 0x^2 + 7x - 14} \\ \underline{6x^5 - 12x^4} \\ 0x^4 + 0x^3 \\ \underline{0x^4 + 0x^3} \\ 0x^3 + 0x^2 \\ \underline{0x^3 + 0x^2} \\ 0x^2 + 7x \\ \underline{0x^2 + 0x} \\ 7x - 14 \end{array}$$

x into $7x$

7 times $x-2$

Subtract

$$\begin{array}{r} 6x^4 + 0x^3 + 0x^2 + 0x + 7 \\ x-2 \overline{) 6x^5 - 12x^4 + 0x^3 + 0x^2 + 7x - 14} \\ \underline{6x^5 - 12x^4} \\ 0x^4 + 0x^3 \\ \underline{0x^4 + 0x^3} \\ 0x^3 + 0x^2 \\ \underline{0x^3 + 0x^2} \\ 0x^2 + 7x \\ \underline{0x^2 + 0x} \\ 7x - 14 \\ \underline{7x - 14} \\ 0 \end{array}$$

525. We were so happy when the house mortgage was paid. It was a 30-year mortgage. (We made 360 monthly payments.)

The first payment paid $100 toward the principal balance. (The total amount owing went down by $100.)

The second payment decreased the principal balance by $107. The third payment decreased the balance by $114. Each monthly payment decreased the principal balance by $7 more than the previous month. By the end of the 360 payments, the balance was zero.

Translation: The 360 payments of 100, 107, 114, 121, ... equaled the original balance on the mortgage. What was that original balance?

First, find the last term.

$a = 100, d = 7, n = 360$ $\ell = a + (n-1)d = 100 + (359)7 = \$2,613$

Then find the sum of all the principal payments.

$$s = \frac{n}{2}(a + \ell) = 180(100 + 2613) = \$488,340.$$

The Complete Solutions and Answers

527. Solve $82^x = 85$

$\log 82^x = \log 85$ Take the log of both sides

$x \log 82 = \log 85$ Birdie Rule

$x = \dfrac{\log 85}{\log 82}$

533. $\dfrac{5x^2 + 18x + 9}{9x^2 + 6x - 8} \div \dfrac{x^2 - 9}{3x^2 - 11x + 6}$

$= \dfrac{(5x^2 + 18x + 9)(3x^2 - 11x + 6)}{(9x^2 + 6x - 8)(x^2 - 9)}$ Invert and multiply

$= \dfrac{(5x + 3)(x + 3)(x - 3)(3x - 2)}{(3x - 2)(3x + 4)(x + 3)(x - 3)}$

$= \dfrac{5x + 3}{3x + 4}$

> Here is the factoring of $5x^2 + 18x + 9$
> $5x^2 + 18x + 9$
> $= 5x^2 + 3x + 15x + 9$ two things that add to $+18x$ and that multiply to $45x^2$
> $= x(5x + 3) + 3(5x + 3)$
> $= (5x + 3)(x + 3)$

539. $\begin{vmatrix} 6 & 8 \\ -1 & 3 \end{vmatrix} = (6)(3) - (-1)(8) = 26$

$\begin{vmatrix} 9 & 2 \\ 0 & -3 \end{vmatrix} = (9)(-3) - (0)(2) = -27$

540. What is the equation of this ellipse?

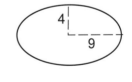

The equation is the easy part: $\dfrac{x^2}{9^2} + \dfrac{y^2}{4^2} = 1$

The English is harder. The length of the **semi-major axis** is 9. The length of the **semi-minor axis** is 4. (The semi-major axis is the longer one. It can be in the x direction or the y direction.) The length of the **major axis** is 18. The length of the **minor axis** is 8.

544–559 The Complete Solutions and Answers

544. Solve $x^2 + 3x + 7 = 0$

Using the quadratic formula
$$x = \frac{-3 \pm \sqrt{9 - (4)(1)(7)}}{2}$$
$$= \frac{-3 \pm \sqrt{-19}}{2}$$
$$= \frac{-3 \pm \sqrt{19}\sqrt{-1}}{2}$$
$$= \frac{-3 \pm \sqrt{19}\,i}{2}$$
OR $\quad \frac{-3}{2} \pm \frac{\sqrt{19}}{2} i$

549. What is the distance between (7, 3) and (5, 8)?

> The distance between two points (x_1, y_1) and (x_2, y_2)
> $$d = \sqrt{(x_2 - x_1)^2 + (y_2 - y_1)^2}$$

$d = \sqrt{(5-7)^2 + (8-3)^2} = \sqrt{4 + 25} = \sqrt{29}$

554. Factor $\quad 40x^4y^6 - 50x^4y^7$

We are looking for the largest expression that will divide evenly into both $40x^4y^6$ and into $-50x^4y^7$.

That expression is $10x^4y^6$. That is the common factor.

$40x^4y^6 - 50x^4y^7$ factors into $10x^4y^6(4 - 5y)$

You can always check a factoring problem by multiplying it out. If you multiply out $10x^4y^6(4 - 5y)$, you will get $40x^4y^6 - 50x^4y^7$.

559. $\quad \dfrac{3}{2x} + \dfrac{2}{x} = \dfrac{7}{10}$

$\dfrac{3(10x)}{2x} + \dfrac{2(10x)}{x} = \dfrac{7(10x)}{10}$ $\quad$ 10x is evenly divisible by 2x, x, and 10

$3(5) + 2(10) = 7x$ $\quad$ All the fractions disappear

$15 + 20 = 7x$

$35 = 7x$

$5 = x$

Checking the x = 5 in the original problem: $\dfrac{3}{10} + \dfrac{2}{5} \stackrel{?}{=} \dfrac{7}{10}$ $\quad$ yes

The Complete Solutions and Answers

564. Resolve into partial fractions $\dfrac{5x^2 - 2x - 69}{(x-7)(x+2)^2}$

This is case 2:
Repeated linear factors.

$$\dfrac{5x^2 - 2x - 69}{(x-7)(x+2)^2} = \dfrac{A}{x-7} + \dfrac{B}{x+2} + \dfrac{C}{(x+2)^2}$$

Multiply each term by $(x-7)(x+2)^2$ to eliminate the denominators

$$5x^2 - 2x - 69 = A(x+2)^2 + B(x-7)(x+2) + C(x-7)$$

The neat shortcut that we used in case 1 (Let x = ...) will not work in case 2 or case 3 or case 4. In this problem, there is no value of x that will eliminate the A and C terms and leave the B term.

Procedure for Cases 2, 3, 4
1. Multiply out the right side of the equation.
2. Equate the corresponding coefficients of each side.

I'll show you how to do that in a tenth of a second from now. The English is harder that than math.

Multiply out the right side of the equation.

$$5x^2 - 2x - 69 = A(x^2 + 4x + 4) + B(x^2 - 5x - 14) + C(x - 7)$$
$$5x^2 - 2x - 69 = Ax^2 + 4Ax + 4A + Bx^2 - 5Bx - 14B + Cx - 7C$$

$$5x^2 - 2x - 69 = (A + B)x^2 + (4A - 5B + C)x + 4A - 14B - 7C$$

This is an equation that is supposed to be true for all x. We are not solving for x. We are looking for the values of A, B, and C. This is something new. We never did this in beginning algebra. That is one reason this is called advanced algebra.

Handy new fact: If an equation is true for all x, then the corresponding coefficients of each power of x must be equal. 5 is the coefficient of x^2 on the left side. A+B is the coefficient of x^2 on the right side. They must be equal. -2 is the coefficient of x on the left. $4A - 5B + C$ is the coefficient of x on the right. -69 is the coefficient of x^0 on the left and $4A - 14B - 7C$ is the coefficient of x^0 on the right. (Recall that x^0 is equal to 1.)

Equating the coefficients of each side.

$$\begin{cases} 5 = A + B \\ -2 = 4A - 5B + C \\ -69 = 4A - 14B - 7C \end{cases}$$

(Continued on next page.)

145

The Complete Solutions and Answers

(564. continued)

Wait a minute! I, your reader, have a question.

Yes. I was expecting that you would interrupt at this point.

That thing that you just wrote—the three equations with the large brace on the left side—that thing looks like three equations with three unknowns.

Yes it does. What was your question?

Isn't that a lot like $\begin{cases} A + B = 5 \\ 4A - 5B + C = -2 \\ 4A - 14B - 7C = -69 \end{cases}$?

Yes it is.

But isn't that a lot like Systems of Equations, which we did in Chapter 5?

Yes it is.

But I thought we were done with that stuff.

If we were "done with that stuff" I never would have taught it in the first place. (⇐ important sentence) Once you solve this system of three equations with three unknowns you have the answer to the partial fractions problem.

I just looked back at my notes. There are three different ways you could solve that. You could do it by the elimination method, by the substitution method, or by Cramer's rule.

Good thought!

What do you mean by "good thought"? I don't like the sound of that.

I was just going to do it by one of the methods, but now that you mention it, doing it by all three methods would be a delight. I'll do the work. You just sit there and watch me sweat out the solution by each of the three methods.

First, I'm going to change the variables to x, y, and z, because I am much more used to them in solving systems of equations.

$$\begin{cases} x + y = 5 \\ 4x - 5y + z = -2 \\ 4x - 14y - 7z = -69 \end{cases}$$

(Continued on next page.)

The Complete Solutions and Answers

(564. continued)

first equation
second equation
third equation
$$\begin{cases} x + y = 5 \\ 4x - 5y + z = -2 \\ 4x - 14y - 7z = -69 \end{cases}$$

Solving by the elimination method

Multiply the second equation by 7 $\qquad 28x - 35y + 7z = -14$
Copy the third equation $\qquad 4x - 14y - 7z = -69$

Add the equations $\qquad 32x - 49y = -83$
Multiply the first equation by 49 $\qquad 49x + 49y = 245$

Add the equations $\qquad 81x = 162$

$\qquad x = 2$

Back substitute into the first equation $\qquad 2 + y = 5$
$\qquad y = 3$

Back substitute $x = 2$ and $y = 3$ into the second equation $\qquad 8 - 15 + z = -2$
$\qquad z = 5$

$$\frac{5x^2 - 2x - 69}{(x - 7)(x + 2)^2} = \frac{2}{x - 7} + \frac{3}{x + 2} + \frac{5}{(x + 2)^2}$$

Solving by the substitution method

Solve the first equation for y $\qquad y = 5 - x$
Substitute $y = 5 - x$ into the second and third equations
$\qquad 4x - 5(5 - x) + z = -2$
$\qquad 4x - 14(5 - x) - 7z = -69$

Simplify the two equations $\qquad 9x + z = 23$
$\qquad 18x - 7z = 1$

Solve the first of these equations for z ($z = 23 - 9x$) and substitute into the other equation
$\qquad 18x - 7(23 - 9x) = 1$

Simplify $\qquad 81x = 162$
$\qquad x = 2$

And now we are back to where we were in solving by the elimination method. Back substituting will yield $y = 3$ and $z = 5$.

(Continued on next page.)

The Complete Solutions and Answers

(564. continued)   first equation
                    second equation
                    third equation
$$\begin{cases} x + y = 5 \\ 4x - 5y + z = -2 \\ 4x - 14y - 7z = -69 \end{cases}$$

Solving by Cramer's rule

$$x = \frac{D_x}{D} = \frac{\begin{vmatrix} 5 & 1 & 0 \\ -2 & -5 & 1 \\ -69 & -14 & -7 \end{vmatrix}}{\begin{vmatrix} 1 & 1 & 0 \\ 4 & -5 & 1 \\ 4 & -14 & -7 \end{vmatrix}}$$

Evaluating D_x by the third column

$$D_x = +0 \begin{vmatrix} -2 & -5 \\ -69 & -14 \end{vmatrix} - 1 \begin{vmatrix} 5 & 1 \\ -69 & -14 \end{vmatrix} + (-7) \begin{vmatrix} 5 & 1 \\ -2 & -5 \end{vmatrix}$$

$$= 0 - (-70 + 69) + (-7)(-25 + 2)$$
$$= 0 + 1 + 161$$
$$= 162$$

Evaluating D by the first row

$$D = 1 \begin{vmatrix} -5 & 1 \\ -14 & -7 \end{vmatrix} - (1) \begin{vmatrix} 4 & 1 \\ 4 & -7 \end{vmatrix} + 0 \begin{vmatrix} 4 & -5 \\ 4 & -14 \end{vmatrix}$$

$$= (35 + 14) - (-28 - 4) + 0$$
$$= 81$$

$$x = \frac{D_x}{D} = \frac{162}{81} = 2$$

And now we are back to where we were in solving by the elimination method.
Back substituting will yield $y = 3$ and $z = 5$.

One nice thing to know is that there are computer programs that can evaluate determinants like $\begin{vmatrix} 5 & 1 & 0 \\ -2 & -5 & 1 \\ -69 & -14 & -7 \end{vmatrix}$ in a single step.
All you do is enter 5, 1, 0, -2, -5, 1, -69, -14, -7 and out will pop 162. They will also do 4×4 determinants and 5×5 and 6×6, etc.

The Complete Solutions and Answers

569. Solve $\dfrac{10x}{3x-5} = \dfrac{1}{x+3}$

    Cross multiply $10x(x+3) = 3x - 5$

    Distributive law $10x^2 + 30x = 3x - 5$

This is a quadratic equation. The easiest way to solve quadratic equations is by factoring—if it factors.

    Set everything equal to zero $10x^2 + 27x + 5 = 0$

This is a trinomial of the form $ax^2 + bx + c$ where $a \neq 1$. Using the "quicker, handy-dandy approach" *(Life of Fred: Beginning Algebra Expanded Edition, page 307)*, we split $+27x$ into two numbers that add to $+27x$ and that multiply to $(10x^2)(+5)$, which is $50x^2$.

    $5x$ and $10x$ multiply to $50x^2$ and add to $15x$. We want $27x$.

    $2x$ and $25x$ multiply to $50x^2$ and add to $27x$. Yes.

    Split the $27x$ into $2x$ and $25x$ $10x^2 + 2x + 25x + 5 = 0$

    Factor by grouping $2x(5x+1) + 5(5x+1) = 0$

                                        $(5x+1)(2x+5) = 0$

    Set each factor equal to zero $5x + 1 = 0$ OR $2x + 5 = 0$

    Solve $x = -1/5$ OR $x = -5/2$

574. Convert $\log_5 9$ into an expression containing only base 10 logs.

$$\log_5 9 = \frac{\log_{10} 9}{\log_{10} 5}$$

579. Given the coordinates of A are (3, 7). The coordinates of B are (15, 28).

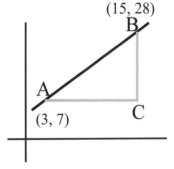

❶ Find the coordinates of C.
B and C have the same x-coordinate.
So C is of the form (15, ?).

A and C have the same y-coordinate.
So C has coordinates (15, 7).

❷ Find the length of BC.
The length of BC is the distance between (15, 28) and (15, 7).
The distance is 21. (details: 28 – 7)

❸ Find the length of AC.
The length of AC is the distance between (3, 7) and (15, 7). It is 12.

❹ Find the slope of AB.   slope $= \dfrac{21}{12}$ or $\dfrac{7}{4}$ if you prefer.

580–590 The Complete Solutions and Answers

580. Let the domain and the codomain be the real numbers.

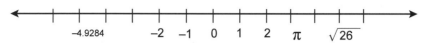

Define f by the rule f(x) = –x.
Is f a function? Is f 1-1? Is f onto?

Under the rule we have, for example:
f(7) = –7
f(–7) = 7
f(0) = 0
f(–$\sqrt{34}$) = $\sqrt{34}$
f(0.95) = –0.95

Every element of the domain has exactly one image. (A number can't have two different negatives.) f is a function.

Are there two different numbers that have the same negative? No. If f(x) = f(y), then it must be true that x = y. f is 1-1.

Is f onto? Yes. Every element of the codomain, (every real number), is the negative of some real number. If you name some real number r, I can find a number x, such that f(x) = r.

584. Explain why division with a domain of the set of all ordered pairs of numbers would *not* be a function.

Not every ordered pair of numbers would have an image in the codomain. For example, (6, 0) would have no meaning. Division by zero is not defined. *Every* element in the domain must have exactly one image in the codomain.

589. What is the degree of each of these terms?

$67x^2y^7$ has degree 9
3.082xyz has degree 3
5y has degree 1
πx has degree 1   π is a number, not a variable.

590. How many functions f are possible if f:{L, M, N} → {P, Q, R, S}?

There are four choices for f(L), either f(L) = P, f(L) = Q, f(L) = R, or f(L) = S.
There are four choices for f(M), either f(M) = P, f(M) = Q, f(M) = R, or f(M) = S.
There are four choices for f(N), either f(N) = P, f(N) = Q, f(N) = R, or f(N) = S.

There are 4×4×4 = 4^3 = 64 possible functions.

The Complete Solutions and Answers

594. I have 40 rabbits. I know that 18 of them like carrots. 25 of them like lettuce. 10 of them like both carrots and lettuce. How many of them like neither?

To begin a Venn diagram problem, often the best place to start is in the inner-most region. I know that 10 of the rabbits like both carrots and lettuce.

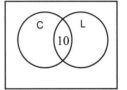

Since 18 like carrots, there must be 8 of them that like carrots but not lettuce.

Since 25 like lettuce, there must be 15 of them that like lettuce and not carrots.

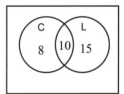

There are 33 (8 + 10 + 15) that like either carrots, lettuce, or both. That leaves 7 (40 − 33) that like neither.

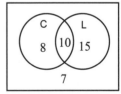

599. $\dfrac{7}{\sqrt{x}+5} = \dfrac{7}{\sqrt{x}+5} \cdot \dfrac{\sqrt{x}-5}{\sqrt{x}-5} = \dfrac{7(\sqrt{x}-5)}{x-25}$ OR $\dfrac{7\sqrt{x}-35}{x-25}$

$\sqrt{x}-5$ is called the **congujate** of $\sqrt{x}+5$.

Just multiplying $\dfrac{7}{\sqrt{x}+5}$ by $\dfrac{\sqrt{x}}{\sqrt{x}}$ won't work. All we will get is $\dfrac{7\sqrt{x}}{x+\sqrt{5x}}$ and there is still a square root in the denominator.

When the denominator is a binomial, such as $\sqrt{x}+5$, you need to use the conjugate.

| 600 | The Complete Solutions and Answers

600. The railing was a mess when I bought my house. Each gallon of Flubber paint takes 2 hours to apply and will cover 100 square feet. Each gallon of Gauss paint takes 1 hour to apply and will cover 200 square feet.

I have at most 10 hours to do the painting and I need to cover at least 1,000 square feet of railing.

Flubber paint costs $12/gallon and Gauss paint costs $90/gallon. How much of each should I use in order to minimize the cost?

Let x = the number of gallons of Flubber paint I will use.
Let y = the number of gallons of Gauss paint I will use.

Reading the English and the previous two lines, I can say that:
It will take 2x + y hours to do the job. $2x + y \leq 10$
The paint will cover 100x + 200y square feet. $100x + 200y \geq 1000$

The cost will be 12x + 90y, which I want to minimize.
The objective function is f(x, y) = 12x + 90y.

Now the easier part. We plot $x \geq 0$, $y \geq 0$, $2x + y \leq 10$, and $100x + 200y \geq 1000$ on the same graph. The 2x + y ≤ 10 will be shaded to the lower left. The 100x + 200y ≥ 1000 will be shaded to the upper right. Where they are all shaded will be a triangle whose vertices are (0, 5), (0, 10), and (10/3, 10/3). I got the (10/3, 10/3) by solving simultaneously 2x + y = 10 and 100x + 200y = 1000.

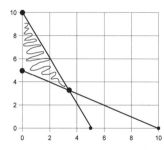

Testing the vertices . . .

f(x, y) = 12x + 90y

   f(0, 5) = $450
   f(0, 10) = $900
   f(10/3, 10/3) = $340 ⇦ the cheapest.

Luckily, these paints come in one-third gallon buckets.

152

The Complete Solutions and Answers   604–619

604. $\log \dfrac{6y + 3}{2w - 1} = \log(6y + 3) - \log(2w - 1)$ the Quotient Rule

609. Suppose you owned 1,000 flamingos. Suppose their population was decreasing by 8% each year because of the neighborhood cats that were attacking them.

How long (to the nearest year) would it be before you had only 700 flamingos?

$$1000(0.92)^x = 700$$

Take the log of both sides $\log 1000(0.92)^x = \log 700$

Product Rule $\log 1000 + \log(0.92)^x = \log 700$

Birdie Rule $\log 1000 + x \log 0.92 = \log 700$

Algebra $x = \dfrac{\log 700 - \log 1000}{\log 0.92}$

Approximating $x = \dfrac{2.845 - 3}{-0.0362} \approx 4.28 \doteq 4$ years

614. Graph $\dfrac{x^2}{7} + \dfrac{y^2}{9} \geq 1$

Draw the fence. (Graph the equality.)

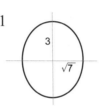

Note:
The ellipse is a solid line because there is $\geq$ in the original inequality and not $>$.

Taste a piece of grass in each field.
(Test one point in each region in the original inequality.)

(0, 0) is inside. $\dfrac{0^2}{7} + \dfrac{0^2}{9} \geq 1$ is false.

(9, 9) is outside. $\dfrac{9^2}{7} + \dfrac{9^2}{9} \geq 1$ is true.

Shade the regions that make the original inequality true.

619. If the domain and codomain of g are the integers { . . . –3, –2, –1, 0, 1, 2, 3, 4 . . .} and g is defined by g(x) = |x|, what is the range of g?

g(–3) = 3   g(–50) = 50   g(0) = 0   g(44) = 44

The set of images of g will be {0, 1, 2, 3, 4, 5 . . .}, which is also known as the whole numbers.

| 624–639 | The Complete Solutions and Answers

624. Translate into an equation: The noise (n) at a party varies directly as the square of the number of people (p) at the party. $n = kp^2$

629. $\sqrt{27x^5} = \sqrt{9x^4}\sqrt{3x} = 3x^2\sqrt{3x}$

634. Factor $12y^2 + y - 6$

Split $+y$ into two things that add to $+y$ and multiply to $-72y^2$ ($= 12y^2$ times -6)

$$= 12y^2 + 9y - 8y - 6$$

Then factor by grouping

$$= 3y(4y + 3) - 2(4y + 3)$$
$$= (4y + 3)(3y - 2)$$

Factor $14x^2 + 35x + 14$

Always look for a common factor first. This problem would be really hard if you didn't do that.

$$= 7(2x^2 + 5x + 2)$$

Split $+5x$ into two things that add to $+5x$ and multiply to $4x^2$ ($= 2x^2$ times 2)

$$= 7(2x^2 + 4x + x + 2)$$

Factor by grouping

$$= 7[2x(x + 2) + (x + 2)]$$
$$= 7(x + 2)(2x + 1)$$

Factor $4w^2 - 23w + 15$

$$= 4w^2 - 20w - 3w + 15$$
$$= 4w(w - 5) - 3(w - 5)$$
$$= (w - 5)(4w - 3)$$

639. Giant rectangle of 7,500 square miles of high-quality farmland in Nevada. Length is 9 miles more than the width. Rounding to the nearest mile, what are the dimensions of the rectangle?

You almost always start a word problem by letting x equal one of the things you are trying to find.
Let x = the width of my wonderful farmland.
Then x + 9 = the length of my wonderful farmland.
Then x(x + 9) = the area of my wonderful farmland, and we know that this is equal to 7,500.
Only after you have written the "Let x equal" and the "Then . . ." statements, do you write the equation.

$x(x + 9) = 7500 \;\rightarrow\; x^2 + 9x - 7500 = 0$ It doesn't factor. Use the quadratic formula.

$$x = \frac{-9 \pm \sqrt{81 - (4)(1)(-7500)}}{2} = \frac{-9 \pm \sqrt{30081}}{2} \doteq 82 \text{ miles (width)}$$

x + 9 = 91 miles (length)   We ignore the negative since we are dealing with length.

The Complete Solutions and Answers

644.
$$a^{10}a^{10} = a^{20}$$
$$(b^7)^3 = b^{21}$$
$$c^8/c^4 = c^4$$

649. Three geese and one boy eat 1.1 pounds of food each day.

Five geese and three boys eat 2.5 pounds of food each day.

$$\begin{cases} 3x + y = 1.1 \\ 5x + 3y = 2.5 \end{cases}$$

Multiply the first equation by −3.
Copy the second equation.

$$\begin{cases} -9x - 3y = -3.3 \\ 5x + 3y = 2.5 \end{cases}$$

Add.

$$-4x = -0.8$$
$$x = 0.2$$

Back substitute $x = 0.2$ into the first equation.

$$3(0.2) + y = 1.1$$
$$y = 0.5$$

One goose eats 0.2 pounds of food per day.
One boy eats 0.5 pounds of food per day.

650. Can you invent a function f with a domain equal to the natural numbers $\{1, 2, 3, 4, 5 \ldots\}$ whose range is $\{2, 4, 6, 8, 10, 12 \ldots\}$?

One possibility is defined by the rule $f(x) = 2x$.
Under this rule:
$$1 \xrightarrow{f} 2$$
$$2 \xrightarrow{f} 4$$
$$3 \xrightarrow{f} 6$$
$$4 \xrightarrow{f} 8 \quad \text{etc.}$$

654. Which of these are graphs of functions?

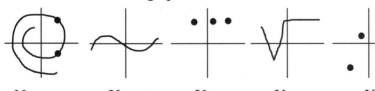

No. Here are two points that are vertically aligned.   Yes.   Yes.   Yes.   Yes.

659–670 The Complete Solutions and Answers

659. $\dfrac{2+3i}{5+6i} = \dfrac{2+3i}{5+6i} \cdot \dfrac{5-6i}{5-6i} = \dfrac{10-12i+15i-18i^2}{25-36i^2}$

$= \dfrac{10+3i+18}{25+36} = \dfrac{28+3i}{61} = \dfrac{28}{61} + \dfrac{3}{61}i$

664. What is the distance between (4, –2) and (–9, 1)?
$d = \sqrt{(-9-4)^2 + (1-(-2))^2} = \sqrt{169+9} = \sqrt{178}$

It was for cases like this that you learned how to add and subtract integers in beginning algebra. 1 – (–2) means 1 subtract –2 which means 1 add +2 which is 3.

665. Describe the largest possible domain for the division function.

We can divide any two numbers as long as the divisor is not zero. The largest possible domain for division is the set of all ordered pairs in which the second coordinate is not zero.

669. Suppose h is a function that maps set C (the domain) <u>onto</u> set D (the codomain). Suppose that the cardinality of set D is 6. What can you say about the cardinality of set C?

[Real life example] Suppose set C is a set of kids. Each kid has a ticket that allows him/her to go on one ride at Dizzyworld. Let h be the function that maps each kid to the ride that the kid chooses. We are told that the number of rides is equal to 6 (the cardinality of D is 6). We know that each of the six rides is chosen by at least one of the kids (h is onto D).

There have to be at least six kids. Otherwise, not all the rides will be chosen. We know all the rides will be chosen because h is onto D.

Therefore, the cardinality of C ≥ 6.

670. What is the degree of each of these polynomials?

$$6x^6 + 43298x - 27xyz$$

The degree of each term: 6   1   3   Degree of the polynomial = 6

$$\pi^2 x^4 + x^5$$

The degree of each term: 4   5   Degree of the polynomial = 5

$$50000x - 0.0003x^3 + \sqrt{7}\,x^2$$

The degree of each term: 1   3   2

Degree of the polynomial = 3

π^2 and $\sqrt{7}$ are numbers, not variables.

The Complete Solutions and Answers | 674–684

674. Solve $\dfrac{8}{x+3} = \dfrac{4}{2x-7}$

    Cross multiplying $8(2x-7) = 4(x+3)$

    Distributive law $16x - 56 = 4x + 12$

    Subtract 4x from both sides $12x - 56 = 12$

    Add 56 to both sides $12x = 68$

    Divide both sides by 12 $x = \dfrac{68}{12} = 5\,{}^2\!/\!_3$

679. $\begin{cases} 3x + y = 5 \\ 7x + 6y = -3 \end{cases}$

Solve the first equation for y $y = 5 - 3x$

Substitute that value of y into
the second equation $7x + 6(5 - 3x) = -3$

                                      $7x + 30 - 18x = -3$

                                      $33 = 11x$

                                      $3 = x$

Back substitute x = 3 into any equation
Using the equation y = 5 – 3x, we obtain $y = 5 - 3(3)$

                                                $y = -4$

684. Solve for y. You do not have to evaluate the determinants.

$$\begin{cases} 3w + 5x + 8y + 2z = 1 \\ 2w + 7x + 3y - 3z = 3 \\ 5w - 3x - 9y + 8z = 4 \\ 7w + 4x + 2y + 5z = -3 \end{cases} \qquad y = \dfrac{D_y}{D}$$

$$y = \dfrac{\begin{vmatrix} 3 & 5 & 1 & 2 \\ 2 & 7 & 3 & -3 \\ 5 & -3 & 4 & 8 \\ 7 & 4 & -3 & 5 \end{vmatrix}}{\begin{vmatrix} 3 & 5 & 8 & 2 \\ 2 & 7 & 3 & -3 \\ 5 & -3 & -9 & 8 \\ 7 & 4 & 2 & 5 \end{vmatrix}}$$

> **In Real Life**
> If you teach math in high school, you will show your students how to solve systems of equations like these that have four equations with four unknowns.
> If you are the mathematician working, say, at an oil refinery, you might be solving a system of 100 equations with 100 unknowns.
> After calculus comes *Life of Fred: Linear Algebra*. There we study all kinds of systems of linear equations. No quadratic equations!

| 689-691 | The Complete Solutions and Answers

689. Put a "☝" under the last non-zero digit or the last gratuitous zero, whichever occurs later.

      9303.0   8000.   47   2020   300.00
        ☝   ☝   ☝   ☝   ☝

690. If f is a function defined by $f(x) = \log_7(x)$, what is $f^{-1}(x)$?

Let's do some experimenting.

    We know that $f(49) = \log_7(49) = 2$, so $f^{-1}(2) = 49 = 7^2$.
    We know that $f(7^3) = \log_7(7^3) = 3$, so $f^{-1}(3) = 7^3$.
    We know that $f(7^{538}) = \log_7(7^{538}) = 538$, so $f^{-1}(538) = 7^{538}$.

Following that pattern, $f^{-1}(x)$ must equal 7^x.

In English: Raising to a power is the inverse of taking a log.

691. Resolve into partial fractions

$$\frac{6x^2 + 17x - 13}{(x+1)(x+3)(x-2)} = \frac{A}{x+1} + \frac{B}{x+3} + \frac{C}{x-2}$$

Eliminate all the denominators by multiplying each term by $(x+1)(x+3)(x-2)$

$$6x^2 + 17x - 13 = A(x+3)(x-2) + B(x+1)(x-2) + C(x+1)(x+3)$$

Let $x = -1$
$$6(-1)^2 + 17(-1) - 13 = A(-1+3)(-1-2)$$
$$-24 = -6A$$
$$4 = A$$

Let $x = -3$
$$6(-3)^2 + 17(-3) - 13 = B(-3+1)(-3-2)$$
$$-10 = 10B$$
$$-1 = B$$

Let $x = 2$
$$6(2)^2 + 17(2) - 13 = C(2+1)(2+3)$$
$$45 = 15C$$
$$3 = C$$

$$\frac{6x^2 + 17x - 13}{(x+1)(x+3)(x-2)} = \frac{4}{x+1} + \frac{-1}{x+3} + \frac{3}{x-2}$$

The Complete Solutions and Answers

694. Solve $\quad 4 = 6^{x-7}$

$\log 4 = \log 6^{x-7}$ $\qquad$ Take the log of both sides

$\log 4 = (x - 7)\log 6$ $\qquad$ Birdie Rule

$\dfrac{\log 4}{\log 6} = x - 7$ $\qquad$ Divide both sides by log 6

$\dfrac{\log 4}{\log 6} + 7 = x$

The exact answer is $\dfrac{\log 4}{\log 6} + 7$. If you were doing engineering and needed to find a decimal approximation, then you would haul out your calculator and do the arithmetic:

$\dfrac{0.6020599913279623904274777 8944899}{0.7781512503836436325087667 9797961} + 7 \quad$ etc.

699. $\dfrac{49a^2 - c^2}{7a^2 + 8ac + c^2}$

$= \dfrac{(7a + c)(7a - c)}{(a + c)(7a + c)}$

$= \dfrac{7a - c}{a + c}$

Factor top:
$49a^2 - c^2 = (7a + c)(7a - c)$
Factor bottom:
$7a^2 + 8ac + c^2$
$\quad = 7a^2 + 7ac + ac + c^2$
$\quad = 7a(a + c) + c(a + c)$
$\quad = (a + c)(7a + c)$

700. Solve by graphing

$\begin{cases} 4x + y = 5 \\ -3x + 2y = 0 \end{cases}$

Putting them both in $\quad \begin{cases} y = -4x + 5 \\ y = \dfrac{3}{2}x + 0 \end{cases}$
$y = mx + b$ form

Graph them on the same set of axes.

The point of intersection looks like about (0.8, 1.3).

The exact answer is (0.875, 1.5).

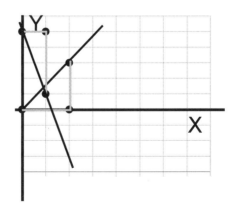

704–719 The Complete Solutions and Answers

704. If a 4-foot Santa weighs 80 pounds, how much would a 6-foot Santa weigh? The weight of an object (w) varies as the cube of the height (h).

Step ①: Find the equation. $w = kh^3$

Step ②: Find the value of k.   If $h = 4$, then $w = 80$.

Substitute that into $w = kh^3$ $80 = k(4^3)$

Divide both sides by 4^3 (which is 64) $\dfrac{80}{64} = k$

$w = kh^3$ now becomes $w = \dfrac{80h^3}{64}$

Step ③: Find w when $h = 6$ $w = \dfrac{80(6^3)}{64}$

Doing the arithmetic $w = 270$

A 6-foot Santa would weigh 270 pounds.

709. $(6x^2 + 48y^{-3x+z} - \sin x + \pi)^0 = 1$ Anything raised to the zero power is equal to one. The only exception is 0^0 which is undefined. That is no big deal since in 40 years of searching I have never been able to find a real-life problem that would result in 0^0. It never seems to come up.

714. If the product of the abscissa and the ordinate of a number is positive, what quadrant or quadrants must it lie in?

If $ab > 0$ for a point (a, b), then either both a and b are positive (in which case it is in quadrant I) or both are negative (in which case it is in quadrant III).

719.

$\dfrac{x}{2} + \dfrac{y}{4} = 1$

$\dfrac{x}{1} + \dfrac{y}{5} = 1$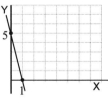

$\dfrac{x}{7} + \dfrac{y}{3} = 1$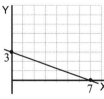

$\dfrac{x}{-3} + \dfrac{y}{-1} = 1$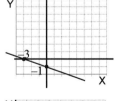

$\dfrac{x}{6} + \dfrac{y}{\pi} = 1$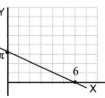

Plotting is just approximate. Just plot (0, 3.1).

$\dfrac{x}{0.7} + \dfrac{y}{5} = 1$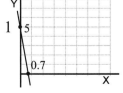

160

The Complete Solutions and Answers

724. Graph $\dfrac{(x-2)^2}{9} - \dfrac{(y-4)^2}{25} = 1$

Draw the egg (ellipse). Center at (2, 4). Semi-major = 5. Semi-minor = 3.

Box the egg (draw rectangle). X out the box.

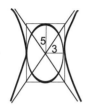

Draw the hyperbola.

Some thoughts about asymptotes . . .
    An asymptote is a line.

A curve will approach its asymptote
if it gets closer and closer to
it as the curve heads farther out. Hyperbolas love their asymptotes.

A parabola is not half of a hyperbola. Parabolas do not have asymptotes.
If you try to draw an asymptote to a parabola, it won't work.
No matter how you try to draw it, the parabola bends away from that line.

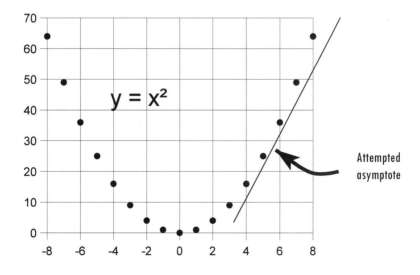

161

729–744 The Complete Solutions and Answers

729. $\dfrac{\sqrt{w}}{\sqrt{xyz}} = \dfrac{\sqrt{w}}{\sqrt{xyz}} \cdot \dfrac{\sqrt{xyz}}{\sqrt{xyz}} = \dfrac{\sqrt{wxyz}}{xyz}$

734. $\log 5x = \log 5 + \log x$ The Product Rule

$\log \dfrac{2}{y} = \log 2 - \log y$ The Quotient Rule

$\log (4 + w)(v) = \log (4 + w) + \log v$

That's as far as we can go. There is no rule that simplifies $\log (m + n)$. Here is one way to think about it:

    The Birdie Rule turns exponents into multiplication: $\log m^n = n \log m$
    The Product Rule turns multiplication into addition: $\log mn = \log m + \log n$
    What in the world could we turn addition into? $\log (m + n) = ?????$

739. Find an approximation for $\log_2 6$, rounding your answer to the hundredth place.

$$\log_2 6 = \dfrac{\log_{10} 6}{\log_{10} 2}$$

$$\approx \dfrac{0.7781512503836436325087667979796\,1}{0.3010299956639811952137388947244\,9}$$

$$= 2.5849625007211561814537389439478$$

$$\doteq 2.58$$

$\approx$ means approximately equal to

$\doteq$ means equal to after rounding

$\log_2 6 = 2.58$ means $2^{2.58}$ will equal 6.

744. Solve $\begin{cases} 11x - 2y = 10 \\ x + 7y = 44 \end{cases}$

Solve the second equation for x $x = 44 - 7y$

Substitute $x = 44 - 7y$ into
the first equation $11(44 - 7y) - 2y = 10$

$484 - 77y - 2y = 10$

$474 = 79y$

$6 = y$

Back substitute $y = 6$ into any equation
Using the equation $x = 44 - 7y$ we obtain $x = 44 - 7(6)$

$x = 2$

The Complete Solutions and Answers | 745–750

745. Invent a function g with domain equal to the whole numbers {0, 1, 2, 3, 4, 5 ...} whose range is the integers { ... –3, –2, –1, 0, 1, 2, 3, 4 ...}.

Here is one way it can be done:

$0 \xrightarrow{g} 0$

$1 \xrightarrow{g} 1$

$2 \xrightarrow{g} -1$

$3 \xrightarrow{g} 2$

$4 \xrightarrow{g} -2$

$5 \xrightarrow{g} 3$

$6 \xrightarrow{g} -3$ etc.

Every integer will be the image of some element in the domain.
In the fancy complicated algebra books, this rule for g could be written:

$$g(x) = \begin{cases} (x+1)/2 & \text{if x is odd} \\ -x/2 & \text{if x is even} \end{cases}$$

749. Which of these are arithmetic series and which are geometric?

$\sum_{i=1}^{6} (9+i) = (9+1) + (9+2) + (9+3)$ etc. $10 + 11 + 12 + 13$ etc.
     This is an arithmetic series. a = 10 and d = 1

$\sum_{i=1}^{3} 7i = 7 + 7 \cdot 2 + 7 \cdot 3 = 7 + 14 + 21$
     This is an arithmetic series. a = 7 and d = 7

$\sum_{i=1}^{9} 4^i = 4 + 4^2 + 4^3 + 4^4 + 4^5$ etc.
     This is a geometric series. a = 4 and r = 4

750. How many possible functions are there for g when we are given g:{1, 2, 3, 4, 5, 6, ..., 99, 100} → {a, b, c}?

There are three choices for g(1): g(1) = a, g(1) = b, or g(1) = c.
There are three choices for g(2): g(2) = a, g(2) = b, or g(2) = c.
There are three choices for g(3): g(3) = a, g(3) = b, or g(3) = c.
There are three choices for g(4): g(4) = a, g(4) = b, or g(4) = c.
There are three choices for g(5): g(5) = a, g(5) = b, or g(5) = c.

By the fundamental principle, there are 3×3×3×3×3×3×3×3×3×3×3 ×3×3×3×3× ... ×3×3×3 (= 3^{100}) ways of creating this function.

754–759 The Complete Solutions and Answers

754. Graph by point-plotting $y = \tan x$ from $x = 0$ to $x = 60$.

If $x = 0$, then $y = 0$ (Details: $y = \tan 0 = 0$) We have the point $(0, 0)$.

If $x = 10$, then $y \doteq 0.18$ (Details: $y = \tan 10 \approx 0.176327 \doteq 0.18$) $(10, 0.18)$

If $x = 20$, then $y \doteq 0.36$ $(20, 0.36)$

and similarly $(30, 0.58)$, $(40, 0.84)$, $(50, 1.2)$, $(60, 1.7)$.

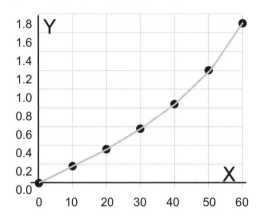

The tangent function ($\tan x$) gets really goofy when x gets near 90.

$\tan 80 \approx 5.7$

$\tan 88 \approx 29.$

$\tan 89.5 \approx 115.$

$\tan 89.9 \approx 573.$

$\tan 89.999 \approx 57,296.$ It is as if $\tan x$ were really ticklish, and 90 was a tickle spot.

759. What is the perimeter of the triangle with vertices $(2, 4)$, $(4, 6)$, and $(5, -7)$?

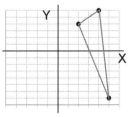

From (2, 4) to (4, 6)

$d = \sqrt{(4-2)^2 + (6-4)^2} = \sqrt{8} = \sqrt{4}\sqrt{2} = 2\sqrt{2}$

From (4, 6) to (5, -7)

$d = \sqrt{(5-4)^2 + (-7-6)^2} = \sqrt{170}$

From (5, -7) to (2, 4)

$d = \sqrt{(2-5)^2 + (4-(-7))^2} = \sqrt{130}$

The perimeter is $2\sqrt{2} + \sqrt{170} + \sqrt{130}$

164

The Complete Solutions and Answers

764. What is the equation of the line with a slope of 5 that passes through the point (4, −17)?

$$m = \frac{y - y_1}{x - x_1} \quad \text{becomes} \quad 5 = \frac{y - (-17)}{x - 4} \quad \text{OR} \quad 5 = \frac{y + 17}{x - 4}$$

769. The minor of 7 in the determinant $\begin{vmatrix} 6 & 7 & 9 \\ 2 & 5 & 1 \\ 0 & 3 & 5 \end{vmatrix}$

is $\begin{vmatrix} 2 & 1 \\ 0 & 5 \end{vmatrix}$

770. In English, tell how you would determine whether the point (7, 13) lies inside the circle $(x - 3)^2 + (y - 5)^2 = 81$.

    This circle has a center at (3, 5) and a radius of 9. All the points inside the circle must be less than 9 units away from the center. Using the distance formula from Chapter 4, I would determine the distance between (3, 5) and (7, 13). If it were less than 9, the point would be inside the circle. (If it were equal to 9, the point would lie on the circle. If the distance were greater than 9, the point would lie outside the circle.)

774. Graph $-\frac{(x + 7)^2}{16} + \frac{(y - 5)^2}{4} > 1$

Draw the graph of the equality. (The fence)
Make it a dashed line since the original was > and not ≥.
The hyperbola is vertical since the plus sign is on the y term.

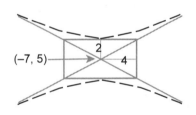

There are three regions to test.
Put (−7, 5) in the original inequality: $-\frac{(-7 + 7)^2}{16} + \frac{(5 - 5)^2}{4} \stackrel{?}{>} 1$ is false.

(−7, 33) is in the top region. $-\frac{(-7 + 7)^2}{16} + \frac{(33 - 5)^2}{4} \stackrel{?}{>} 1$ is true.

(−7, −55) is in the bottom region. $-\frac{(-7 + 7)^2}{16} + \frac{(-55 - 5)^2}{4} \stackrel{?}{>} 1$ is true.

Shade the upper and lower regions
that make the original inequality true.

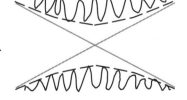

| | The Complete Solutions and Answers |

779. $\dfrac{25x^4 + 12x^5 + 7x^3 + 8 + 15x^2 + 14x}{3x + 1}$

Arrange in descending powers

$$3x + 1 \overline{\smash{\big)}\, 12x^5 + 25x^4 + 7x^3 + 15x^2 + 14x + 8}$$

3x into $12x^5$
$4x^4$ times $3x + 1$

$$\begin{array}{r} 4x^4 \\ 3x + 1 \overline{\smash{\big)}\, 12x^5 + 25x^4 + 7x^3 + 15x^2 + 14x + 8} \\ 12x^5 + 4x^4 \end{array}$$

Subtract
Bring down $7x^3$

$$\begin{array}{r} 4x^4 \; + 7x^3 \\ 3x + 1 \overline{\smash{\big)}\, 12x^5 + 25x^4 + 7x^3 + 15x^2 + 14x + 8} \\ 12x^5 + 4x^4 \\ 21x^4 + 7x^3 \end{array}$$

3x into $21x^4$
$7x^3$ times $3x + 1$

$$\begin{array}{r} 4x^4 \; + 7x^3 \\ 3x + 1 \overline{\smash{\big)}\, 12x^5 + 25x^4 + 7x^3 + 15x^2 + 14x + 8} \\ 12x^5 + 4x^4 \\ 21x^4 + 7x^3 \\ 21x^4 + 7x^3 \end{array}$$

Subtract
Bring down $15x^2 + 14x$

$$\begin{array}{r} 4x^4 \; + 7x^3 \\ 3x + 1 \overline{\smash{\big)}\, 12x^5 + 25x^4 + 7x^3 + 15x^2 + 14x + 8} \\ 12x^5 + 4x^4 \\ 21x^4 + 7x^3 \\ \underline{21x^4 + 7x^3} \\ + 15x^2 + 14x \end{array}$$

3x into $15x^2$
$5x$ times $3x + 1$

$$\begin{array}{r} 4x^4 \; + 7x^3 + 5x \\ 3x + 1 \overline{\smash{\big)}\, 12x^5 + 25x^4 + 7x^3 + 15x^2 + 14x + 8} \\ 12x^5 + 4x^4 \\ 21x^4 + 7x^3 \\ \underline{21x^4 + 7x^3} \\ + 15x^2 + 14x \\ + 15x^2 \; + 5x \end{array}$$

The Complete Solutions and Answers

Subtract
Bring down +8

$$\begin{array}{r} 4x^4 + 7x^3 + 5x \\ 3x + 1 \overline{\smash{)}\, 12x^5 + 25x^4 + 7x^3 + 15x^2 + 14x + 8} \\ \underline{12x^5 + 4x^4} \\ 21x^4 + 7x^3 \\ \underline{21x^4 + 7x^3} \\ + 15x^2 + 14x \\ \underline{+ 15x^2 + 5x} \\ 9x + 8 \end{array}$$

3x into 9x
3 times 3x + 1
Subtract

$$\begin{array}{r} 4x^4 + 7x^3 + 5x + 3 \\ 3x + 1 \overline{\smash{)}\, 12x^5 + 25x^4 + 7x^3 + 15x^2 + 14x + 8} \\ \underline{12x^5 + 4x^4} \\ 21x^4 + 7x^3 \\ \underline{21x^4 + 7x^3} \\ + 15x^2 + 14x \\ \underline{+ 15x^2 + 5x} \\ 9x + 8 \\ \underline{9x + 3} \\ 5 \end{array}$$

Put the remainder up
as a fraction

$$\begin{array}{r} 4x^4 + 7x^3 + 5x + 3 + \tfrac{5}{3x+1} \\ 3x + 1 \overline{\smash{)}\, 12x^5 + 25x^4 + 7x^3 + 15x^2 + 14x + 8} \\ \underline{12x^5 + 4x^4} \\ 21x^4 + 7x^3 \\ \underline{21x^4 + 7x^3} \\ + 15x^2 + 14x \\ \underline{+ 15x^2 + 5x} \\ 9x + 8 \\ \underline{9x + 3} \\ 5 \end{array}$$

    Once you get used to it, long division of polynomials gets to be very routine. You divide, multiply, subtract, and bring down followed by divide, multiply, subtract, and bring down followed by divide, multiply, subtract, and bring down followed by divide, multiply, subtract, and bring down followed by divide, multiply, subtract, and bring down till you get to the end. As they say, it is long, but not hard.

784–794 The Complete Solutions and Answers

784. Factor $50x^2 - 200y^2 = 50(x^2 - 4y^2) = 50(x + 2y)(x - 2y)$

Always look for a common factor first.

$x^2 + 8x + 16 = (x + 4)(x + 4)$ or $(x + 4)^2$

$15x^2 - 2x - 8$

To find two things that add to $-2x$ and multiply to $-120x^2$ ($= 15x^2$ times -8)

$= 15x^2 + 10x - 12x - 8$

Then factor by grouping

$= 5x(3x + 2) - 4(3x + 2)$
$= (3x + 2)(5x - 4)$

789. $6x^2 = 54$
$x^2 = 9$
$x = \pm 3$

$y^2 - 27 = 0$
$y^2 = 27$
$y = \pm\sqrt{27} = \pm\sqrt{9}\sqrt{3} = \pm 3\sqrt{3}$

$z^2 + 6z = 2(3z + 50)$
$z^2 + 6z = 6z + 100$
$z^2 = 100$
$z = \pm 10$

When some people write z in algebra, they cross their z so that it looks like this: ƶ. That prevents getting 2 mixed up with z.

794. There are two ways that you can tell if a pair of equations is inconsistent:

A) What do their graphs look like?

Their graphs are parallel lines.

B) What do you get when you try to solve them by the elimination method?

Both the x terms and the y terms will disappear at the same time and you get something that is never true, such as $0 = 73$.

Two inconsistent equations are like two people that are always fighting. If a pair of numbers satisfies one of the two equations, it is certain that they will not satisfy the other equation.

Think of something like: $\begin{cases} x + y = 4 \\ x + y = 5 \end{cases}$

Two numbers that add to 4 will never add to 5.

The Complete Solutions and Answers

799. Redoing the previous graph with a "slight" change. Graph $y = x^2 - 4x + 12$ from $x = -8$ to $x = 12$. Please plot at least five points.

Again, I'm going to be silly and plot zillions of points.

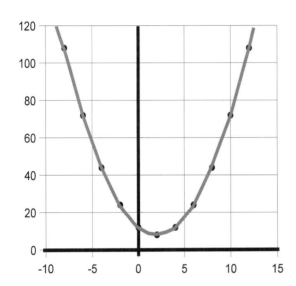

x	y
-8	108
-6	72
-4	44
-2	24
0	12
2	8
4	12
6	24
8	44
10	72
12	108

804. Graph $4x^2 - 25y^2 = 100$

Put into standard form. (Divide by 100.) $\qquad \dfrac{x^2}{5^2} - \dfrac{y^2}{2^2} = 1$

Graph the egg (ellipse).

Box the egg and X it out.

Draw the hyperbola.

Most professional hyperbola artists
skip drawing the egg and just draw the egg box.

169

The Complete Solutions and Answers

809. Graph $y < 2x^2 + 3$

Draw the fence. (Graph the equality.)
Point-plotting $y = 2x^2 + 3$
$x = 0, y = 3$
$x = 1, y = 5$
$x = -1, y = 5$
$x = \pm 2, y = 11$ (I can do this $\pm$ since they both give the same answer.)
$x = \pm 3, 21$

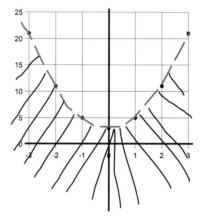

Test a point in each region.

$(0, 0)$ is below the curve. $0 \stackrel{?}{<} 2(0)^2 + 3$ is true.

$(0, 40)$ is above the curve. $40 \stackrel{?}{<} 2(0)^2 + 3$ is false.

Shade the area(s) that test "true."

814. How many terms are in the sequence 23, 26, 29, 32, ..., 1949?

$\ell = a + (n-1)d$ becomes
$$1949 = 23 + (n-1)3$$
$$1949 = 23 + 3n - 3$$
$$1949 = 20 + 3n$$
$$1929 = 3n$$
$$643 = n$$

815. In Chapter 7 we introduced the notation f:A → B. This meant a function whose name is f, whose domain is set A, and whose codomain is set B.

How many possible functions are there when f:{1, 2, 3} → {7, 8}?

I have two choices for f(1): either f(1) = 7 or f(1) = 8.
I have two choices for f(2): either f(2) = 7 or f(2) = 8.
I have two choices for f(3): either f(3) = 7 or f(3) = 8.

By the fundamental principle, there are 2×2×2 (= 8) ways to create a function f:{1, 2, 3} → {7, 8}.

The Complete Solutions and Answers

819. Find the mean and median averages of 3.5, 4.3, 4.2

To find the mean average of three numbers, add them and divide the sum by three. $3.5 + 4.3 + 4.2 = 12$ $12 \div 3 = 4$

To find the median average, arrange the numbers from smallest to largest and pick the one in the middle. 3.5, 4.2, 4.3

The median average is 4.2.

824. Simplify $\sqrt{300}$ $\sqrt{300} = \sqrt{100}\sqrt{3} = 10\sqrt{3}$

829. $z + \dfrac{1}{z+2} = \dfrac{6z+1}{6}$

$\dfrac{z}{1} + \dfrac{1}{z+2} = \dfrac{6z+1}{6}$ Make each term into a fraction

$\dfrac{z(6)(z+2)}{1} + \dfrac{(6)(z+2)}{z+2} = \dfrac{6(z+2)(6z+1)}{6}$

$6(z+2)$ is evenly divisible by $z+2$ and by 6

$6z(z+2) + 6 = (z+2)(6z+1)$ and all the fractions disappear

$6z^2 + 12z + 6 = 6z^2 + z + 12z + 2$

$4 = z$

Checking $z = 4$ in the original equation $4 + \dfrac{1}{6} \stackrel{?}{=} \dfrac{25}{6}$ yes

834. Place $4x^2 + 25y^2 = 100$ into the standard form for ellipses. Then graph the ellipse.

$$4x^2 + 25y^2 = 100$$

Divide each term by 100 $\dfrac{x^2}{25} + \dfrac{y^2}{4} = 1$

Looking at the equation in standard form, we can tell that the length of the semi-major axis is 5, and it is in the horizontal direction. The length of the semi-minor axis is 2, and it is in the vertical direction.

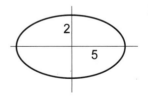

More English: The vertices of an ellipse are located at the ends of the major axis. In this case, they are (5, 0) and (–5, 0).

The **center** of an ellipse is the point halfway between the vertices. (*Vertices* is the plural of vertex.) The center of this ellipse is the origin.

835–839 | The Complete Solutions and Answers

835. The domain is the set of whole numbers $\{0, 1, 2, 3, 4, \ldots\}$, and the codomain is the set of natural numbers $\{1, 2, 3, 4, \ldots\}$. Define f by the rule $f(x) = 2x$. Is f a function?

If f is going to be a function, then each element of $\{0, 1, 2, 3, \ldots\}$ must have exactly one image in $\{1, 2, 3, \ldots\}$.

However, what does $f(0)$ equal? It has no image in the natural numbers. f is not a function.

839. Unadd $\dfrac{15x^3 - 5x^2 - 76x - 20}{(x^2 + 7x + 2)(3x^2 - 4)}$

$$\frac{15x^3 - 5x^2 - 76x - 20}{(x^2 + 7x + 2)(3x^2 - 4)} = \frac{Ax + B}{x^2 + 7x + 2} + \frac{Cx + D}{3x^2 - 4}$$

Multiply each term by $(x^2 + 7x + 2)(3x^2 - 4)$ to eliminate the denominators

$$15x^3 - 5x^2 - 76x - 20 = (Ax + B)(3x^2 - 4) + (Cx + D)(x^2 + 7x + 2)$$
$$15x^3 - 5x^2 - 76x - 20 = 3Ax^3 - 4Ax + 3Bx^2 - 4B + Cx^3 + 7Cx^2 + 2Cx + Dx^2 + 7Dx + 2D$$

$$15x^3 - 5x^2 - 76x - 20 = (3A + C)x^3 + (3B + 7C + D)x^2 + (-4A + 2C + 7D)x + (-4B + 2D)$$

Equating the coefficients of each side

$$\begin{cases} 15 = 3A + C \\ -5 = 3B + 7C + D \\ -76 = -4A + 2C + 7D \\ -20 = -4B + 2D \end{cases}$$

Switching the sides of the equations

$$\begin{cases} 3A \quad\ \ + C \qquad\qquad = 15 & \text{first equation} \\ \qquad 3B + 7C + D = -5 & \text{second equation} \\ -4A \qquad + 2C + 7D = -76 & \text{third equation} \\ \qquad -4B \qquad + 2D = -20 & \text{fourth equation} \end{cases}$$

The second equation times -7 and add to the third. The second equation times -2 and add to the fourth. These two new equations and the first equation are three equations in three unknowns. Using Cramer's rule for C on these three equations yields $C = 0$. Back substitute into the first equation gives $A = 5$. Using $A = 5$ in the third equation gives $D = -8$. Using $D = -8$ in the fourth equation gives $B = 1$.

$$\frac{15x^3 - 5x^2 - 76x - 20}{(x^2 + 7x + 2)(3x^2 - 4)} = \frac{5x + 1}{x^2 + 7x + 2} + \frac{-8}{3x^2 - 4}$$

The Complete Solutions and Answers | 844–859

844. $y^{1/2} = \sqrt{y}$
$27^{1/3} = 3$ since $3^3 = 27$
$\sqrt{2}\sqrt{8} = \sqrt{16} = 4$

849. $i^{102} = i^{100}i^2 = (i^4)^{25}i^2 = 1^{25}i^2 = i^2 = -1$
$(6 + 2i)^2 = (6 + 2i)(6 + 2i) = 36 + 12i + 12i + 4i^2 = 32 + 24i$
$(-7i)(-8i) = 56i^2 = -56$

854. Actually, a number in scientific notation is in the form $d \times 10^n$ where $1 \le d < 10$ isn't quite true. Notice that -513.7 in scientific notation is -5.137×10^2. Rewrite the rule to take into account negative numbers.

First approach: A number in scientific notation is in the form $d \times 10^n$ where either $1 \le d < 10$ or $-10 < d \le -1$.

Second approach: A number in scientific notation is in the form $d \times 10^n$ where the absolute value of d is greater than or equal to one and less than ten.

Third approach: A number in scientific notation is in the form $d \times 10^n$ where $1 \le |d| < 10$.

There may be other approaches to defining scientific notation. There is one number that cannot be put into scientific notation. That number is zero. Every other number that can be expressed as a decimal can be written as $d \times 10^n$ where $1 \le |d| < 10$.

859. During the summer, the number of weeds in my garden grows by 5% each day. How many days will it take for the number of weeds in my garden to triple?

$(1.05)^x = 3$

Take the log of both sides $\log (1.05)^x = \log 3$

Birdie Rule $x \log 1.05 = \log 3$

Divide both sides by log 1.05 $x = \dfrac{\log 3}{\log 1.05}$

$\dfrac{\log 3}{\log 1.05}$ is the exact answer. If we approximate the logarithms, we get:

$x \approx \dfrac{0.477}{0.0212} \doteq 23$ days

864–874 The Complete Solutions and Answers

864. Given the coordinates of A are (8, 13) and the coordinates of B are (11, 19), find the slope of the line through A and B.

❶ Find the coordinates of C.
B and C have the same x-coordinate.
So C is of the form (11, ?).

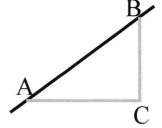

A and C have the same y-coordinate.
So C has coordinates (11, 13).

❷ Find the length of BC.
The length of BC is the distance between (11, 13) and (11, 19).
The distance is 6. (details: 19 − 13)

❸ Find the length of AC.
The length of AC is the distance between (8, 13) and (11, 13). It is 3.

❹ Find the slope of AB.   slope = $\frac{6}{3}$ OR 2 if you prefer.

869. What is the equation of the line with a slope of ⅔ that intercepts the y-axis at (0, 7)?

$$y = \tfrac{2}{3}x + 7$$ Don't you wish all questions were this easy!

874. Evaluate $\begin{vmatrix} 4 & 6 & -7 \\ 0 & 3 & 5 \\ -2 & 2 & 8 \end{vmatrix}$

This is called expansion by minors.

First, pick any row or any column. You will get the same answer regardless of which one you choose, but sometimes your work will be less if you pick one with nice numbers (like zero) in it.

I'll pick the second row.

Second, the answer is ■(0)(the minor of 0) + ■(3)(the minor of 3) + ■(5)(the minor of 5)
where ■ is either + or − according to this chart: + − +
                                                    − + −
                                                    + − +

Using the second row: $-(0)\begin{vmatrix} 6 & -7 \\ 2 & 8 \end{vmatrix} + (3)\begin{vmatrix} 4 & -7 \\ -2 & 8 \end{vmatrix} - (5)\begin{vmatrix} 4 & 6 \\ -2 & 2 \end{vmatrix}$

= 0 + 3(32 − 14) − 5(8 + 12)
=   3(18) − 5(20) = −46

The Complete Solutions and Answers

875. Graph $\dfrac{(x-5)^2}{9} - \dfrac{(y+2)^2}{3} \le 1$

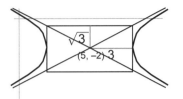

Draw the graph of the equality. (The fence)
Make it a solid line since the original was $\le$ rather than $<$.

There are three regions to test.
Put $(5, -2)$ in the original inequality: $\dfrac{(5-5)^2}{9} - \dfrac{(-2+2)^2}{3} \overset{?}{\le} 1$ is true.

$(-90, -2)$ is in the left region. $\dfrac{(-90-5)^2}{9} - \dfrac{(-2+2)^2}{3} \overset{?}{\le} 1$

roughly   1003 $-$ 0 $\overset{?}{\le}$ 1 is false.

$(70, -2)$ is in the right region. $\dfrac{(70-5)^2}{9} - \dfrac{(-2+2)^2}{3} \overset{?}{\le} 1$

roughly   469 $-$ 0 $\overset{?}{\le}$ 1 is false.

Shade the region that makes
the original inequality true.

879. $A = \begin{pmatrix} 3 & 5 & 11 & 2 \\ 4 & 7 & 8 & 6 \\ 1 & 0 & 23 & 7 \end{pmatrix}$ ⇦ pairs of children's shoes
⇦ pairs of women's shoes
⇦ pairs of men's shoes

pink red black brown

$C = \begin{pmatrix} 5 & 1 & 22 & 3 \\ 0 & 2 & 14 & 5 \\ 0 & 0 & 38 & 4 \end{pmatrix}$ ⇦ pairs of children's shoes
⇦ pairs of women's shoes
⇦ pairs of men's shoes

pink red black brown

$A + C = \begin{pmatrix} 8 & 6 & 33 & 5 \\ 4 & 9 & 22 & 11 \\ 1 & 0 & 61 & 11 \end{pmatrix}$ I told you it wasn't hard.

884–894 The Complete Solutions and Answers

884. The time (t) it takes to drive to my house from your house varies inversely as the speed (s) you are driving. At 50 miles per hour, it would take 8 hours to get there. How long would it take at 40 miles per hour?

Step ①: Find the equation. $t = \dfrac{k}{s}$

Step ②: Find the value of k. If $s = 50$, then $t = 8$.

Substituting that into $t = \dfrac{k}{s}$ we get

$$8 = \dfrac{k}{50}$$

Multiply both sides by 50

$$400 = k$$

$t = \dfrac{k}{s}$ now becomes

$$t = \dfrac{400}{s}$$

Step ③: Find t when $s = 40$.

$$t = \dfrac{400}{40}$$

$$t = 10$$

At 40 mph it will take 10 hours to drive to my house.

889. Graph $y = \dfrac{-2}{3}x + 8$

The slope $m = \dfrac{-2}{3}$ and the y-intercept is at (0, 8).

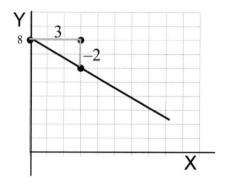

894. $\dfrac{2x^2 + 7x - 23}{x^2 + 11x + 30} - \dfrac{x - 1}{x + 6}$

$\dfrac{2x^2 + 7x - 23}{x^2 + 11x + 30} + \dfrac{-x + 1}{x + 6}$ The first step in a subtraction problem is to turn it into an addition problem.

$\dfrac{a}{b} - \dfrac{c}{d}$ becomes $\dfrac{a}{b} + \dfrac{-c}{d}$

$= \dfrac{2x^2 + 7x - 23}{(x + 5)(x + 6)} + \dfrac{-x + 1}{x + 6}$ Factor the denominator(s)

$= \dfrac{2x^2 + 7x - 23}{(x + 5)(x + 6)} + \dfrac{(-x + 1)(x + 5)}{(x + 6)(x + 5)}$ Interior decoration

$= \dfrac{2x^2 + 7x - 23 - x^2 - 4x + 5}{(x + 5)(x + 6)}$

$= \dfrac{x^2 + 3x - 18}{(x + 5)(x + 6)} = \dfrac{(x + 6)(x - 3)}{(x + 5)(x + 6)} = \dfrac{x - 3}{x + 5}$

The Complete Solutions and Answers

895. There were a lot of rocks in my backyard that I needed to remove. I had two options. I could either use a hammer and wheelbarrow or use dynamite and a truck.

With the hammer and wheelbarrow, my cost would be $1/day and I would get injured about 0.2 times each day.

With the dynamite and a truck, my cost would be $50/day and I would get injured about 0.1 times each day.

I had $817 to spend on the project and I could experience 5 injuries at most.

I can remove 4 tons of rocks each day using a hammer and wheelbarrow and 20 tons using dynamite and a truck. What is my best course of action in order to remove as much rock as possible?

My backyard

My goal is to remove as much rock as possible, but my question is how should I allocate my time between hammer/wheelbarrow and dynamite/truck.

Let x = number of days I use hammer and wheelbarrow.
Let y = number of days I use dynamite and a truck.

Using these two lines and the statement of the problem, I can say that
x + 50y is the total cost and 0.2x + 0.1y is the total injuries.

Since I can spend a maximum of $817 and experience a maximum of 5 injuries
$$x + 50y \leq 817$$
$$0.2x + 0.1y \leq 5$$

4x + 20y is the total number of tons I can remove. I want to maximize this. f(x, y) = 4x + 20y is the objective function.

Plot x ≥ 0, y ≥ 0, x + 50y ≤ 817, and 0.2x + 0.1y ≤ 5.

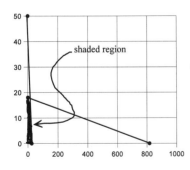

Solving x + 50y = 817 and
0.2x + 0.1y = 5 simultaneously,
I get x = 17 and y = 16.

Testing the four vertices in f(x, y) = 4x + 20y
f(0, 0) = 0 tons of rocks
f(0, 16.34) = 326.8 tons of rocks
f(17, 16) = 388 tons of rocks ⇦ winner
f(25, 0) = 100 tons of rocks

Seventeen days with a hammer and wheelbarrow and 16 days with dynamite and a truck will allow me to haul away 388 tons.

> 899–909 **The Complete Solutions and Answers**

899. Solve $\quad 4x^2 + 12x + 5 = 0$

$$x = \frac{-12 \pm \sqrt{144 - (4)(4)(5)}}{8}$$

$$x = \frac{-12 \pm \sqrt{64}}{8}$$

$$x = \frac{-12 \pm 8}{8} = \frac{-4}{8} \text{ OR } \frac{-20}{8} = -\tfrac{1}{2} \text{ OR } -2\tfrac{1}{2}$$

Wait a minute! Whenever the $b^2 - 4ac$ is a perfect square (such as 4, 9, 16, 25, 36, 49, etc.), then the final answers will not have a square root in it. That means that the original equation could have been solved by factoring.

> In short, if $b^2 - 4ac$ is a perfect square, you didn't need to use the quadratic formula.

In the language of math, $b^2 - 4ac$ is called the discriminant.

Here is the solution of the original equation by factoring: $\quad 4x^2 + 12x + 5 = 0$

$$(2x + 5)(2x + 1) = 0$$
$$2x + 5 = 0 \quad \text{OR} \quad 2x + 1 = 0$$
$$x = -2\tfrac{1}{2} \quad \text{OR} \quad x = -\tfrac{1}{2}$$

Solve $\quad 6x^2 + x + 2 = 0$

$$x = \frac{-1 \pm \sqrt{1 - (4)(6)(2)}}{12}$$

$$x = \frac{-1 \pm \sqrt{-47}}{12}$$

$$x = \frac{-1 \pm \sqrt{-1}\sqrt{47}}{12} = \frac{-1 \pm i\sqrt{47}}{12}$$

904. $d^{-4}/d^{-6} = d^{-4-(-6)} = d^{-4+(+6)} = d^2$
$(m^7 n^2)^3 = m^{21} n^6$
$a^0 = 1$

909. Place $\dfrac{-2 + i}{2 - i}$ in the form $a + bi$.

$$\frac{-2 + i}{2 - i} = \frac{-2 + i}{2 - i} \cdot \frac{2 + i}{2 + i} = \frac{-4 - 2i + 2i + i^2}{4 - i^2}$$

$$= \frac{-4 - 1}{4 + 1} = -1 \quad \text{OR} \quad -1 + 0i$$

If we had factored -1 out of the numerator of the original problem, we would have had

$\dfrac{(-1)(2 - i)}{2 - i}$ and after canceling $= -1$.

The Complete Solutions and Answers

914. Graph $36x^2 + 16y^2 = 576$. What are the coordinates of the vertices?

The first step is to put the equation into standard form for the ellipse.

$$36x^2 + 16y^2 = 576$$

Divide both sides by (36)(16)

$$\frac{x^2}{16} + \frac{y^2}{36} = 1$$

$$\frac{x^2}{4^2} + \frac{y^2}{6^2} = 1$$

The semi-minor axis has a length of 4 and is in the horizontal direction. The semi-major axis has a length of 6 and is in the vertical direction.

The vertices are at the ends of the semi-major axes. (0, 6) and (0, –6)

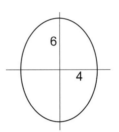

Here is the ellipse with the same shape whose center is at (443, 398):

$$\frac{(x-443)^2}{4^2} + \frac{(y-398)^2}{6^2} = 1$$

915. We will graph a parabola that opens to the right. Graph $x = 8y^2$ from $y = -3$ to $y = 3$.

Again, I'm going to be silly and plot zillions of points. I name a value for y first, and then compute the corresponding value for x.

x	y
72	-3
32	-2
8	-1
0	0
8	1
32	2
72	3

179

| 919–929 | The Complete Solutions and Answers |

919. Solve $\dfrac{x-9}{5} = \dfrac{-4}{x}$

   Cross multiply $x(x-9) = -20$
   Distributive law $x^2 - 9x = -20$

This is a quadratic equation.
The easiest way to solve it is by factoring—if it factors.

   Set everything equal to zero $x^2 - 9x + 20 = 0$

Look for two numbers that add to –9 and multiply to +20.

   It factors $(x-4)(x-5) = 0$

When two things multiply to zero, one of them must be zero.

   Set each factor equal to zero $x - 4 = 0$ OR $x - 5 = 0$
   Solve each equation $x = 4$ OR $x = 5$

Let's check those answers in the original problem.
Checking $x = 4$

   $\dfrac{4-9}{5} \stackrel{?}{=} \dfrac{-4}{4}$ becomes $\dfrac{-5}{5} \stackrel{?}{=} \dfrac{-4}{4}$ yes

Checking $x = 5$

   $\dfrac{5-9}{5} \stackrel{?}{=} \dfrac{-4}{5}$ becomes $\dfrac{-4}{5} \stackrel{?}{=} \dfrac{-4}{5}$ yes

924. What is the point at which $y = 8.3x + 2.705$ intercepts the y-axis?

  $y = 8.3x + 2.705$ is in the form $y = mx + b$, so we can tell instantly that it intercepts the y-axis at the point (0, 2.705).

929. Factor $3x^2 - 13x + 12$
     $= 3x^2 - 4x - 9x + 12$
     $= x(3x - 4) - 3(3x - 4)$
     $= (3x - 4)(x - 3)$

Factor $9y^2 + 21y + 10$
     $= 9y^2 + 6y + 15y + 10$
     $= 3y(3y + 2) + 5(3y + 2)$
     $= (3y + 2)(3y + 5)$

Factor $40x^2 - 40x + 10$
     $= 10(4x^2 - 4x + 1)$ Always look for a common factor first!
     $= 10(4x^2 - 2x - 2x + 1)$
     $= 10[2x(2x - 1) - (2x - 1)]$
     $= 10(2x - 1)(2x - 1)$ or $10(2x - 1)^2$

The Complete Solutions and Answers

934. Underline the first non-zero digit, put a "☜" under the last non-zero digit or the last gratuitous zero, whichever occurs later, and state how many significant digits are in each number.

    0.00<u>3</u>004 ☜   4 significant digits

    <u>1</u>70 ☜   2 significant digits

    <u>2</u>.0003 ☜   5 significant digits

    <u>3</u>9.50 ☜   4 significant digits

    000<u>5</u>.006 ☜   4 significant digits

    <u>8</u>,000,000 ☜   1 significant digit

939. The point (6, 9) is in which quadrant?
Answer: The point (6, 9) is in quadrant I.

    What is the abscissa of (6, 9)?
Answer: The abscissa is the x-coordinate. The abscissa of (6, 9) is 6.

    If (a, b) is in quadrant II, what can you say about b?
Answer: Points in quadrant II have negative x-coordinates and positive y-coordinates. b > 0

944. Given points (x_1, y_1) and (x_2, y_2) on line ℓ. Find the slope of ℓ.

❶ Find the coordinates of C.
    (x_2, y_1) B and C have the same abscissas.
                     A and C have the same ordinates.

❷ Find the length of BC.
    $y_2 - y_1$ From (x_2, y_1) to (x_2, y_2)

❸ Find the length of AC.
    $x_2 - x_1$ From (x_1, y_1) to (x_2, y_1)

❹ Find the slope of AB.
    $\dfrac{y_2 - y_1}{x_2 - x_1}$

> The slope formula given two points (x_1, y_1) and (x_2, y_2)
> $$m = \dfrac{y_2 - y_1}{x_2 - x_1}$$

947–959 | The Complete Solutions and Answers

947. What is the equation of the line that passes through (–4, 7) and (3, 10)?

$$\frac{y - y_1}{x - x_1} = \frac{y_2 - y_1}{x_2 - x_1} \quad \text{becomes} \quad \frac{y - 7}{x + 4} = \frac{10 - 7}{3 + 4}$$

$$\text{OR} \quad \frac{y - 7}{x + 4} = \frac{3}{7} \quad \text{OR} \quad 3x - 7y = -61$$

950. Solve $-5 = \sqrt{14w + 3}$

There is no solution. A $\sqrt{}$ can never equal a negative number.
(See principal square roots in Life of Fred: Beginning Algebra Expanded Edition, page 380.)

953. $\log 67 + \log xyz = \log 67xyz \quad$ since $\log m + \log n = \log mn$

956. Factor
$y^2 - 12y + 11 = (y - 1)(y - 11)$
$w^2 - 2w - 15 = (w - 5)(w + 3)$
$x^2 + 3x + 40 \quad$ does not factor
$5z^2 + 45z + 70 = 5(z^2 + 9z + 14) = 5(z + 2)(z + 7)$

Always look for a common factor first.

959. $\dfrac{4}{6 - x} + \dfrac{2}{3x} = \dfrac{4}{3}$

$$\frac{4(3x)(6 - x)}{6 - x} + \frac{2(3x)(6 - x)}{3x} = \frac{4(3x)(6 - x)}{3}$$

Notice that I used $(3x)(6 - x)$ and not $3(3x)(6 - x)$. All three denominators will divide evenly into $(3x)(6 - x)$, so why make extra work by using $3(3x)(6 - x)$?

$$4(3x) + 2(6 - x) = 4x(6 - x) \quad \text{No fractions!}$$
$$12x + 12 - 2x = 24x - 4x^2$$
$$4x^2 - 14x + 12 = 0$$
$$2x^2 - 7x + 6 = 0 \quad \text{I made my life easier by dividing both sides of the equation by 2.}$$
$$(x - 2)(2x - 3) = 0$$
$$x - 2 = 0 \quad \text{OR} \quad 2x - 3 = 0$$
$$x = 2 \quad \text{OR} \quad x = \frac{3}{2}$$

Checking x = 2

$\dfrac{4}{6 - 2} + \dfrac{2}{6} \stackrel{?}{=} \dfrac{4}{3} \quad$ yes

Checking x = $\dfrac{3}{2}$

$\dfrac{4}{6 - 1.5} + \dfrac{2}{3(1.5)} \stackrel{?}{=} \dfrac{4}{3}$

after much arithmetic → $\dfrac{8}{9} + \dfrac{4}{9} \stackrel{?}{=} \dfrac{4}{3} \quad$ yes

The Complete Solutions and Answers 960–963

960. Graph $-4x^2 + 25y^2 = 100$

$$-\frac{x^2}{5^2} + \frac{y^2}{2^2} = 1$$

Previous problem (#804) was $4x^2 - 25y^2 = 100$

Algebra in the previous problem gave

$$\frac{x^2}{5^2} - \frac{y^2}{2^2} = 1$$

The graph was

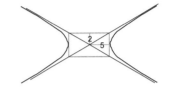

So the graph in this problem will be

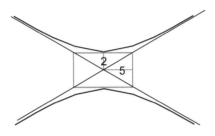

962. Define function h as $\{(3, 9), (5, 7), (-9, 8), (6, 7)\}$.

What is the domain of h? The domain of h is the set of first coordinates. The domain is $\{3, 5, -9, 6\}$.

What is the range of h? The set of images of function h is the set of second coordinates: $\{9, 7, 8\}$. (In listing a function inside of braces, it is considered a "misspelling" to list the same element twice.)

Is h 1-1? Since $5 \xrightarrow{h} 7$ and $6 \xrightarrow{h} 7$, it is not one-to-one.

963. Find the sum of $\sum_{i=1}^{\infty} 2(0.4^i) = 2(0.4) + 2(0.4)^2 + 2(0.4)^3 + \ldots$

$a = 2(0.4)$ which is 0.8

$r = 0.4$ and we note that $-1 < 0.4 < 1$.

$$s = \frac{a}{1-r} = \frac{0.8}{1-0.4} = \frac{0.8}{0.6} = 1\frac{1}{3}$$

| 965–971 | **The Complete Solutions and Answers**

965. Given lines ℓ_1 and ℓ_2 are perpendicular, and they have slopes m_1 and m_2.

If $m_1 = \frac{2}{5}$ m_2 equals $-\frac{5}{2}$ since $(\frac{2}{5})(-\frac{5}{2}) = -1$

If $m_1 = 4$, m_2 equals $-\frac{1}{4}$ since $(4)(-\frac{1}{4}) = -1$

If $m_1 = -0.01$ m_2 equals 100 since $(-0.01)(100) = -1$

    In the language of math, two lines are perpendicular if their slopes are the negative reciprocals of each other. $\frac{a}{b}$ and $\frac{-b}{a}$

968. Find the length of the hypotenuse.

By the Pythagorean theorem

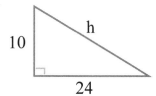

$$10^2 + 24^2 = h^2$$
$$100 + 576 = h^2$$
$$676 = h^2$$
$$\pm 26 = h$$ Since the length cannot be negative, the final answer is 26.

971. There are two ways that you can tell if a pair of equations is dependent:

    A) What do their graphs look like?

Their graphs are the same line drawn twice.

    B) What do you get when you try to solve them by the elimination method?

Both the x terms and the y terms will disappear at the same time and you get something that is always true, such as $0 = 0$.

    Two dependent equations are like twins that are always in agreement. If a pair of numbers satisfies one of the two equations, it is certain that it will satisfy the other equation.

    Think of something like:
$$\begin{cases} x + y = 4 \\ 3x + 3y = 12 \end{cases}$$

    Two numbers that add to 4, then three times the first number plus three times the second number will always equal 12.

The Complete Solutions and Answers

974. Example A: {(8, 8), (9, 2), (2, 9)} Is this a function?
8 has exactly one image in the codomain.
9 has exactly one image in the codomain.
2 has exactly one image in the codomain.
 This is the definition of a function.

Example B: {(1, 2), (2, 4), (3, 6), (3, 8)} Is this a function?
3 has two images in the codomain. It is mapped to both 6 and 8.
This is not a function.

Example C: {(5, a), (b, a), (7, c), (9, d)} Is this function 1-1?
Both 5 and b are mapped to a. This function is not 1-1.

Example D: {(M, L), (L, P), (P, Q), (Q, R)} Is this function 1-1?
Since no second coordinate is mentioned more than once, this must be a 1-1 function. (← Hint: This is another way of determining whether a function is 1-1.)
Another hint: If no first coordinate is mentioned more than once, the set of ordered pairs must be a function.

975. Just do the first step.

$$\frac{5x^4 + 2x + 88}{(2x+7)^2(x-3)} = \frac{A}{2x+7} + \frac{B}{(2x+7)^2} + \frac{C}{x-3}$$

$$\frac{6x^3 + 5x + 34}{(9x-1)(x^2+3)^2} = \frac{A}{9x-1} + \frac{Bx+C}{x^2+3} + \frac{Dx+E}{(x^2+3)^2}$$

$$\frac{7}{x(2x^2+3)^3} = \frac{A}{x} + \frac{Bx+C}{2x^2+3} + \frac{Dx+E}{(2x^2+3)^2} + \frac{Fx+G}{(2x^2+3)^3}$$

977. How many possible functions are there for f if f:A→B where A has 6 members and B has 8 members and f is 1-1. (A function is 1-1 if no two members of the domain have the same image.)

There are 8 possible choices for the image of the first member of A.
There are 7 possible choices for the image of the second member of A because f is 1-1. The image of the second member of A can't be the same as the image of the first member of A.
There are 6 possible choices for the image of the third member of A.
 By the fundamental principle, there are 8×7×6×5×4×3 (= 20,160) possible functions.

| 980–989 | **The Complete Solutions and Answers**

980. Put into scientific notation:
$600 = 6 \times 10^2$
$0.005 = 5 \times 10^{-3}$
$30.08 = 3.008 \times 10$
$2 = 2 \times 10^0$
$10^6 = 1 \times 10^6$

983. What are the coordinates of the origin? **Answer:** The origin is (0, 0).

Name a point (a, b) where a > 0 that does not lie in either quadrant I or quadrant IV.
Answer: Any point on the positive x-axis, such as (7, 0) or (π, 0), will work.

Is it possible for the graph of a straight line to lie in exactly two quadrants?
Answer: Yes. Vertical lines, such as x = 5 or x = –3, will only intersect two quadrants. Horizontal lines, such as y = 8, will only intersect two quadrants.

986. $\dfrac{2}{x-5} + \dfrac{8}{5-x}$

$= \dfrac{2}{x-5} + \dfrac{8(-1)}{(5-x)(-1)}$

$= \dfrac{2}{x-5} + \dfrac{-8}{x-5}$ *Once you have seen this trick, you can amaze your friends at parties.*

$= \dfrac{-6}{x-5}$ OR $\dfrac{6}{5-x}$

989. Solve $4x - 7 = -2x^2$ / $(5x + 3)(2x - 5) = -24$

The first step is to put the equation into standard form: $ax^2 + bx + c = 0$.

$2x^2 + 4x - 7 = 0$ $10x^2 - 25x + 6x - 15 = -24$
                       $10x^2 - 19x + 9 = 0$

$x = \dfrac{-4 \pm \sqrt{16 - (4)(2)(-7)}}{4}$ $x = \dfrac{19 \pm \sqrt{361 - (4)(10)(9)}}{20}$

$x = \dfrac{-4 \pm \sqrt{72}}{4} = \dfrac{-4 \pm \sqrt{36}\sqrt{2}}{4}$ $x = \dfrac{19 \pm 1}{20}$

$x = \dfrac{-4 \pm 6\sqrt{2}}{4} = \dfrac{2(-2 \pm 3\sqrt{2})}{4}$ $x = 1$ OR $x = \dfrac{9}{10}$

$x = \dfrac{-2 \pm 3\sqrt{2}}{2}$

The Complete Solutions and Answers

992. Solve $\begin{cases} 3x + 5y = 41 \\ 2x - 6y = -38 \end{cases}$

We have two choices: ❶ Multiply the first equation by 6 and the second equation by 5 and add. This will eliminate the y terms. Or ❷ multiply the first equation by 2 and the second equation by −3 and add. This will eliminate the x terms.

I choose ❷: $\begin{cases} 6x + 10y = 82 \\ -6x + 18y = 114 \end{cases}$

Add the two equations $\qquad 28y = 196$

$$y = 7$$

Substitute $y = 7$ into the first equation $\qquad 3x + 5(7) = 41$

$$3x = 6$$
$$x = 2$$

993. Continuing the previous problem (#770), determine whether the point (7, 13) lies inside the circle $(x - 3)^2 + (y - 5)^2 = 81$.

First, find the distance between (7, 13) and the center of the circle (3, 5) using the distance formula $d = \sqrt{(x_2 - x_1)^2 + (y_2 - y_1)^2}$

$d = \sqrt{(7-3)^2 + (13-5)^2} = \sqrt{16 + 64} = \sqrt{80}$
The radius of the circle is 9.
$\sqrt{80}$ is less than 9 since $\sqrt{81}$ is equal to 9.
The point is inside the circle.

995. Let g be a function whose domain is {a, b, c, d} and whose codomain is {a, b, c, d, e, f }. Could g be 1-1? Could g be onto?

g could be 1-1. Here is one possible example:
$\quad a \xrightarrow{g} d$
$\quad b \xrightarrow{g} f$
$\quad c \xrightarrow{g} a$
$\quad d \xrightarrow{g} c$

g could not be onto. The four members of the domain could hit, at most, four members of the codomain. At least two members of the codomain will not be images of elements of the domain.

998. How many 6-letter "words" are there? There are 52 possible first letters, 52 possible second letters, etc. Using the fundamental principle, there are $52 \times 52 \times 52 \times 52 \times 52 \times 52 = 52^6$ possible "words."

If you like that multiplied out, it's 19,770,609,664 "words."

| 1001–1007 | The Complete Solutions and Answers |

1001. $8 + \sqrt{w + 13} = 12$

First, isolate the radical on one side of the equation $\quad \sqrt{w + 13} = 4$
Second, square both sides $\quad w + 13 = 16$
$\quad w = 3$

Third, check your answer in the original equation
$8 + \sqrt{3 + 13} \stackrel{?}{=} 12$
$8 + \sqrt{16} \stackrel{?}{=} 12$
$8 + 4 \stackrel{?}{=} 12 \quad$ yes

$w = 3$ is the solution to $8 + \sqrt{w + 13} = 12$.

1004. You put $287 into a savings account. It grows by 5% each year. That means that after one year you would have 287(1.05).
After two years you would have $287(1.05)(1.05) = 287(1.05)^2$.
After three years, $287(1.05)^3$.
Approximately how many years would it take for you to have a million dollars in your account?

Translation: $\quad 287(1.05)^x = 1{,}000{,}000$

Take the log of both sides $\quad \log 287(1.05)^x = \log 1{,}000{,}000$

Product Rule $\quad \log 287 + \log (1.05)^x = \log 1{,}000{,}000$

$\log_{10} 1{,}000{,}000$
$= \log_{10} 10^6 = 6 \quad \log 287 + \log (1.05)^x = 6$

Birdie Rule $\quad \log 287 + x \log 1.05 = 6$
Subtract log 287 $\quad x \log 1.05 = 6 - \log 287$

Divide both sides by log (1.05) $\quad x = \dfrac{6 - \log 287}{\log 1.05}$

The exact answer is $\dfrac{6 - \log 287}{\log 1.05}$ but we are asked for an approximation.

Using a calculator $\dfrac{6 - 2.45788189673399232522162 81085608}{0.02118929906993807279350526712326}$

$\doteq 167$ years

1007. Translate into an equation: The chance (C) that you will become a dean at KITTENS University varies directly as the square of the amount of money (m) that you donate to the President's Fund and inversely as the number of enemies (e) that you have made.

$$C = \frac{km^2}{e}$$

188

The Complete Solutions and Answers | 1010–1013

1010. Solve $\dfrac{x-2}{3} = \dfrac{-1}{x-6}$

Cross multiply $\qquad\qquad\qquad (x-2)(x-6) = -3$

Multiply the two binomials $\qquad x^2 - 6x - 2x + 12 = -3$

To multiply two binomials $(a + b)(c + d)$

first multiply the a times the c and the d $\qquad (a + b)(c + d) \qquad ac + ad$

and then multiply the b times the c and the d $\qquad (a + b)(c + d) \qquad bc + bd$

Combine like terms $\qquad\qquad\qquad x^2 - 8x + 12 = -3$

This is a quadratic equation. The easiest way to solve quadratic equations is by factoring—if it factors.

Set everything equal to zero $\qquad x^2 - 8x + 15 = 0$

Look for two numbers that add to –8 and multiply to +15.

It factors $\qquad\qquad\qquad (x - 3)(x - 5) = 0$

If two things multiply to zero, one of them must be zero.

$\qquad\qquad\qquad x - 3 = 0 \quad \text{OR} \quad x - 5 = 0$

Solve each equation $\qquad\qquad x = 3 \quad \text{OR} \quad x = 5$

1013. $\dfrac{x+10}{4x-24} + \dfrac{4}{x^2-8x+12} = \dfrac{5}{x-6}$

$\dfrac{x+10}{4(x-6)} + \dfrac{4}{(x-2)(x-6)} = \dfrac{5}{x-6}$ factor denominators

$\dfrac{(x+10)(4)(x-6)(x-2)}{4(x-6)} + \dfrac{4(4)(x-6)(x-2)}{(x-2)(x-6)} = \dfrac{5(4)(x-6)(x-2)}{x-6}$

If you hadn't factored the denominators, you might have used $(4x - 24)(x^2 - 8x + 12)(x - 6)$, which would have created an equation that would be almost impossible to solve.

$(x + 10)(x - 2) + 16 = 20(x - 2) \qquad$ No fractions!

$x^2 + 8x - 20 + 16 = 20x - 40$

$x^2 - 12x + 36 = 0$

$(x - 6)(x - 6) = 0 \qquad$ Solve by factoring

$x - 6 = 0 \qquad$ No need to write: "…OR $x - 6 = 0$"

$x = 6$

Checking $x = 6$ in the original equation: $\dfrac{16}{4(6) - 24} \ldots$ Stop! There is a zero in the denominator. The original equation has no solution.

| 1016–1022 | **The Complete Solutions and Answers**

1016. Graph $16(x-3)^2 + (y-5)^2 = 16$.
What are the coordinates of the center and of the vertices?

$$16(x-3)^2 + (y-5)^2 = 16$$

Divide each term by 16.
$$(x-3)^2 + \frac{(y-5)^2}{16} = 1$$

$$\frac{(x-3)^2}{1} + \frac{(y-5)^2}{4^2} = 1$$

The center of the ellipse is (3, 5).
The length of the semi-major axis is 4 in the vertical direction.
The length of the semi-minor axis is 1 in the horizontal direction.

Since the center is at (3, 5), and the length of the semi-major axis is 4, the vertices will be at (3, 1) and (3, 9).

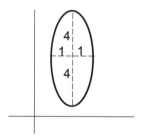

1019. Solve $x^2 = -1$

    Take the square root of both sides. $x = \pm\sqrt{-1}$
    The definition of i. $x = \pm i$

Recall that if you have a pure quadratic such as $x^2 = 7$, you solve it by taking the square root of both sides of the equation and putting a $\pm$ sign on the number. $x = \pm\sqrt{7}$

1022. My favorite pizza place is always open. It is 6 a.m. and I need my breakfast pizza. (Is there a better way to start the day?) I run from my house toward the pizzeria and cover half of the distance in 8 minutes. I continue running and cover half of the remaining distance in 4 minutes. And cover half the remaining distance in 2 minutes and so on. What time will it be when I get there?

Adding $8 + 4 + 2 + 1 + \frac{1}{2} + \frac{1}{4} + \frac{1}{8} + \ldots$ is finding the sum of an infinite geometric series whose first term is 8 and whose common ratio is 1/2. a = 8 and r = 0.5

$$s = \frac{a}{1-r} = \frac{8}{1-0.5} = 16$$

It will be 6:16 a.m. when I get there. We have added an infinite number of positive numbers and have a finite answer. Math can be weird.

The Complete Solutions and Answers | 1025

1025. While vacationing in sunny Sandeneyes, I visited their famous Pottery Shop.

    On the top row were round pots that cost $3 each, weighed 5 pounds, and could hold 2 liters of root beer.

    On the bottom row were tall, skinny pots that cost $6 each, weighed 15 pounds, and could hold 5 liters of root beer.

    My budget was $30 for buying pottery. I could carry at most 60 pounds. How many of each should I buy so that they would hold as much root beer as possible?

Here is a cute little table . . .

	cost	weight	volume
x of the round pots	$3	5 pounds	2 liters
y of the tall pots	$6	15 pounds	5 liters

The constraints:

My budget was $30 for buying pottery. $3x + 6y \le 30$

I could carry at most 60 pounds. $5x + 15y \le 60$

The objective function:

To hold as much root beer as possible.   to maximize $f(x, y) = 2x + 5y$

Point-plotting: $(10, 0)$ and $(0, 5)$ are on $3x + 6x = 30$
                           $(0, 4)$ and $(12, 0)$ are on $5x + 15y = 60$

The intersection of these two lines is at $(6, 2)$. I used Cramer's rule, even though the elimination method would probably be easier.

$$x = \frac{D_x}{D} = \frac{\begin{vmatrix} 30 & 6 \\ 60 & 15 \end{vmatrix}}{\begin{vmatrix} 3 & 6 \\ 5 & 15 \end{vmatrix}} = \frac{(30)(15)-(60)(6)}{(3)(15)-(5)(6)} = \frac{90}{15} = 6$$

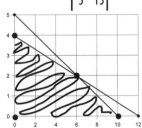

Testing the four vertices in $f(x, y) = 2x + 5y$

    $f(0, 4) = 20$ liters
    $f(6, 2) = 22$ liters ⇦ maximum volume
    $f(10, 0) = 20$ liters
    $f(0, 0) = 0$ liters

I should buy 6 round pots and 2 tall ones.

| 1028–1034 | The Complete Solutions and Answers

1028. Solve $\dfrac{-1}{5x^2 - 2x + 3} = \dfrac{3}{20x - 17}$

    Cross multiply $-(20x - 17) = 3(5x^2 - 2x + 3)$
    Distributive law $-20x + 17 = 15x^2 - 6x + 9$

This is a quadratic equation. The easiest way to solve quadratic equations is by factoring—if it factors.

    Set everything equal to zero $0 = 15x^2 + 14x - 8$

$15x^2 + 14x - 8$ is a trinomial of the form $ax^2 + bx + c$ where $a \neq 1$.

Using the "quicker, handy-dandy approach" (from *Life of Fred: Beginning Algebra Expanded Edition*, page 307)
we split $+14x$ into two numbers that add to $+14x$ and that multiply to $(15x^2)(-8)$, which is $-120x^2$.

    $2x$ and $-60x$ multiply to $-120x^2$ and add to $-58x$. They are too far apart.
    $8x$ and $-15x$ multiply to $-120x^2$ and add to $-7x$. They are too close to each other.
    $6x$ and $-20x$ multiply to $-120x^2$ and add to $-14x$. We want $+14x$. Switch the signs.
    $-6x$ and $+20x$ multiply to $-120x^2$ and add to $+14x$. ☺

    Split the $+14x$ into $-6x$ and $20x$ $0 = 15x^2 - 6x + 20x - 8$
    Factor by grouping $0 = 3x(5x - 2) + 4(5x - 2)$
                $0 = (5x - 2)(3x + 4)$

If two things multiply to zero, one of them must be zero.

    $5x - 2 = 0$ OR $3x + 4 = 0$
    $x = \dfrac{2}{5}$ OR $x = -\dfrac{4}{3}$

1031. Solve $y^2 + 5 = 3$
    This is a pure quadratic equation.
    Put it into the form $y^2 = $ a number. $y^2 = -2$
    Take the square root of both sides. $y = \pm\sqrt{-2}$
    Since $\sqrt{ab} = \sqrt{a}\sqrt{b}$ $y = \pm\sqrt{2}\sqrt{-1}$
    Definition of i $y = \pm\sqrt{2}\,i$

1034. $\log x^7 = 7 \log x$ The Birdie Rule

$\log (w + \pi)^6 = 6 \log (w + \pi)$ The Birdie Rule   This cannot be simplified any further.

$\log \dfrac{x - 3y}{x - 3y} = \log (x - 3y) - \log (x - 3y) = 0$ The Quotient Rule

The Complete Solutions and Answers 1037–1043

1037. You have a million dollars. Each year the government taxes away 20% of what you own. *Approximately* how many years would it take for you to have $1000?

After one year you would have 1,000,000(0.8).
After x years you would have 1,000,000(0.8)^x.
To solve $\quad 10^6(0.8)^x = 10^3$

$\log 10^6(0.8)^x = \log 10^3$	Take the log of both sides
$\log 10^6 + \log (0.8)^x = \log 10^3$	Product Rule
$6 + \log (0.8)^x = 3$	$\log_b b^m = m$
$\log (0.8)^x = -3$	Subtract 6 from both sides
$x \log 0.8 = -3$	Birdie Rule
$x = \dfrac{-3}{\log 0.8}$	Divide both sides by log 0.8

$x \approx \dfrac{-3}{-0.09691} \approx 30.9 \doteq 31$ years

1040. Without point-plotting, graph $y = \dfrac{2}{5}x + 4$.

① The y-intercept is (0, 4).
② At that intercept draw a triangle with rise of 2 and run of 5.
③ Draw the line.

This is a zillion times faster than point-plotting. If the equation is in y = mx + b form, it is bang, bang, bang! You dot the y-intercept. You draw the triangle for the slope. You slash in the line.

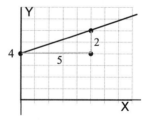

On a slow day, you can (with practice) graph one of these every five seconds. Intercept/Triangle/Slash. Intercept/Triangle/Slash. Intercept/Triangle/Slash. Intercept/Triangle/Slash. Intercept/Triangle/Slash. Intercept/Triangle/Slash. That's six graphs in a half minute. Point-plotting will graph any equation, but if you have y = mx + b, there is no need to fool around with the "If x = ..., then y = ..." stuff.

1043. Find the mean and median averages of 3, 9, 5, 6, 12.

To find the mean average of five numbers, add them up and divide the sum by five. $3 + 9 + 5 + 6 + 12 = 35 \quad\quad 35 \div 5 = 7$

To find the median average, arrange the numbers from smallest to largest and pick the one in the middle. 3, 5, 6, 9, 12 The median average is 6.

| 1046–1052 | The Complete Solutions and Answers

1046. $\sqrt{70-y} - 2 = y$

    First, isolate the radical. $\sqrt{70-y} = y + 2$
    Second, square both sides. $70 - y = y^2 + 4y + 4$
    Solving by factoring. $0 = y^2 + 5y - 66$

looking for two numbers that add to +5 and
that multiply to –66

$$0 = (y-6)(y+11)$$
$$y - 6 = 0 \quad \text{OR} \quad y + 11 = 0$$
$$y = 6 \quad \text{OR} \quad y = -11$$

    Third, check each answer in the original equation.

checking y = 6

$\sqrt{70-6} - 2 \stackrel{?}{=} 6$
$\sqrt{64} - 2 \stackrel{?}{=} 6$
$8 - 2 \stackrel{?}{=} 6$ yes

y = 6 is a solution to the equation.

checking y = –11

$\sqrt{70-(-11)} - 2 \stackrel{?}{=} -11$
$\sqrt{81} \stackrel{?}{=} -9$ no

We don't have to go any further. A $\sqrt{}$ can never equal a negative number. See principal square roots in Life of Fred: Beginning Algebra Expanded Edition, page 380.)

1049. Solve for z using Cramer's rule. You do not have to evaluate the determinants.

$$\begin{cases} 3x + 2y - 35z = 2 \\ 9x - 8y + 67z = 3 \\ 2x + 3y + 4z = 4 \end{cases} \qquad z = \frac{D_z}{D}$$

$$z = \frac{\begin{vmatrix} 3 & 2 & 2 \\ 9 & -8 & 3 \\ 2 & 3 & 4 \end{vmatrix}}{\begin{vmatrix} 3 & 2 & -35 \\ 9 & -8 & 67 \\ 2 & 3 & 4 \end{vmatrix}}$$

1052. What are the coordinates of the center and of the vertices of the hyperbola $\frac{(x-7)^2}{100} - \frac{(y+23)^2}{144} = 1$?

The center is (7, –23).
The vertices (the black dots) are at (7 ± 10, –23).

194

The Complete Solutions and Answers   1055–1058

1055. $(x^4y^6)^4 = x^{16}y^{24}$
$z^{-3}/z^{-6} = z^{-3-(-6)} = z^{-3+(+6)} = z^3$
$w^{-100} = 1/w^{100}$

1058. Joe owns 37 boats and ships.
  23 of them can go on the ocean ($\mathscr{O}$).
  20 have sails ($\mathscr{S}$).
  22 have a picture on Fred painted on the hull ($\mathscr{F}$).
  15 are ocean-going and have a picture of Fred on the hull.
  13 have sails and a picture of Fred on the hull.
  10 are ocean-going and have sails.
  7 are ocean-going, have sails, and have a picture of Fred.
How many are not ocean-going, have no sails, and no picture of Fred?

ocean-going with sails

Start with the inner-most region.

7 are ocean-going, have sails, and have a picture of Fred

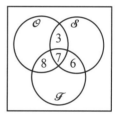
not ocean-going with Fred

8 since 15 are ocean-going with Fred. (15 – 7)
6 since 13 are sails with Fred. (13 – 7)
3 since 10 are ocean-going with sails. (10 – 7)

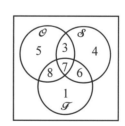

5 ocean-going only. (23 – 8 – 7 – 3)
4 sails only. (20 – 3 – 7 – 6)
1 Fred only. (22 – 8 – 7 – 6)

3 are not ocean-going, have no sails, and do not have a picture of Fred on the hull.
(37 – 5 – 3 – 7 – 8 – 6 – 4 – 1)

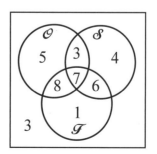

195

| 1061–1075 | The Complete Solutions and Answers

1061. $\log_{\sqrt{6}} 6 = 2$ since $(\sqrt{6})^2 = 6$
That is the definition of square root.

$\log_{0.1} 100 = -2$ since $(0.1)^{-2} = (\frac{1}{10})^{-2} = 10^2 = 100$

$\log_2 2^6 = 6$ $\quad$ $\log_2 2^6$ asks the question, "To what power do you raise 2 in order to get 2^6?"
Even your baby brother knows the answer to that question.

1064. In an evening of studying, the number of facts (F) learned varies directly as the square root of the number of hours (h) spent.
$$F = k\sqrt{h}$$

1067. Graph $\frac{x}{8} + \frac{y}{3} = 1$

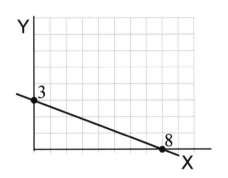

1072. $20 - \frac{x+6}{x+3}$

$= \frac{20}{1} - \frac{x+6}{x+3}$ $\quad$ If they aren't fractions, make them fractions.

$= \frac{20}{1} + \frac{-x-6}{x+3}$ $\quad$ Every subtraction problem is turned into an addition problem.

$= \frac{20(x+3)}{(x+3)} + \frac{-x-6}{x+3}$ $\quad$ Interior decoration

$= \frac{20x + 60 - x - 6}{x+3}$

$= \frac{19x + 54}{x+3}$ $\quad$ This fraction cannot be simplified.

1075. Examples B, C, and D are the ones that can be multiplied together. Two matrices can be multiplied if the number of columns in the first matrix equals the number of rows in the second. In Example B, we are multiplying a 2×3 times a 3×1 matrix. The answer will be a 2×1 matrix.

The Complete Solutions and Answers | 1077

1077. $\dfrac{43}{\sqrt{w}-7} = \dfrac{43}{\sqrt{w}-7} \cdot \dfrac{\sqrt{w}+7}{\sqrt{w}+7} = \dfrac{43(\sqrt{w}+7)}{w-49}$

$\sqrt{w}+7$ is the conjugate of $\sqrt{w}-7$.

There is one special reason why we learn to rationalize the denominator when it is a binomial. This reason is rarely mentioned in advanced algebra books. Math teachers keep it a secret. If I tell you now, please don't tell anyone where you learned this.

THE SECRET: In the first semester of calculus, we will learn to take derivatives (whatever that means). We will do it by the delta-process (whatever that means). Δ is the Greek letter delta.

When we use the delta-process to find the derivative of $y = \sqrt{6x}$, we will come to the point $\dfrac{\Delta y}{\Delta x} = \dfrac{\sqrt{6x+6\Delta x} - \sqrt{6x}}{\Delta x}$ and we will need to rationalize the numerator!

We will multiply top and bottom by the conjugate $\dfrac{\sqrt{6x+6\Delta x} + \sqrt{6x}}{\sqrt{6x+6\Delta x} + \sqrt{6x}}$ and get

$\dfrac{\Delta y}{\Delta x} = \dfrac{6x + 6\Delta x - 6x}{\Delta x(\sqrt{6x+6\Delta x} + \sqrt{6x})} = \dfrac{6\Delta x}{\Delta x(\sqrt{6x+6\Delta x} + \sqrt{6x})}$

and the Δx cancels nicely: $\dfrac{6}{\sqrt{6x+6\Delta x} + \sqrt{6x}}$

Now, you may be asking, "Why in the world did we want to get rid of the Δx in the denominator?"

The answer is even more embarrassing than admitting that you learned to rationalize a binomial denominator because some day you would need to rationalize a binomial numerator.

We needed to get rid of the Δx in the denominator, because in calculus we are going to take $\dfrac{\Delta y}{\Delta x}$ and we are going to let Δx get infinitely close to zero.

✔ You have been told a zillion times that you can't have a denominator equal to zero.

✔ When you were studying arithmetic, you were told that you can't subtract 7 from 2.

✔ When you were studying sets, you learned that the cardinality of $\{☎, ✢, ♦, ▭\}$ is four. You were told that there is no number that is the cardinality of $\{1, 2, 3, 4, 5, 6, \ldots\}$.

Today you know that we can subtract 7 from 2.   (Answer $= -5$)

In a year or two, when you study calculus, we will take fractions and push the denominators infinitely close to zero.

As a math major in college, you will learn that the cardinality of $\{1, 2, 3, 4, 5, 6, \ldots\}$ is a number—a number bigger than any number you could ever think of. It's called $\aleph_0$. (aleph null)

And then, of course, there is $\aleph_1$ and $\aleph_2$. In fact, there are an infinite number of these numbers. And this is all just the beginning. There are numbers beyond $\aleph_{\aleph_0}$

1082–1087 | **The Complete Solutions and Answers**

1082. Draw a Venn diagram of the union of the set of all log cabins (L) and all buildings (B).

   Since every log cabin is a building, the set of all log cabins is a subset of the set of all buildings, $L \subset B$.

   The set of everything that is either a log cabin or a building is the set of all buildings. $L \cup B \subset B$.

1087. Which of these is nonsense?

   $\log_1 8$ asks the question, "To what power do you raise 1 in order to get an answer of 8?"

   In symbols, $1^? = 8$.

   The difficulty is that one to any power is equal to one.

   One is not a good base for logarithms.

   $\log_0 298$ asks the question, "To what power do you raise 0 in order to get an answer of 298?"

   In symbols, $0^? = 298$.

   The difficulty is that zero to any power is equal to zero.

   Zero is not a good base for logarithms.

   $\log_\pi 6$ asks the question, "To what power do I raise π in order to get an answer of 6?"

   In symbols $\pi^? = 6$.

   That is hard, but not impossible. In fact, in the next section of this book we will solve $\pi^x = 6$. Here's how it is done:

Take the log of both sides of the equation	$\log \pi^x = \log 6$
Birdie Rule	$x \log \pi = \log 6$
Divide both sides by $\log \pi$	$x = \dfrac{\log 6}{\log \pi}$

   If I use my calculator *to get an approximation,*
   $$x = \frac{0.7781512503836436325087667979796l}{0.4971498726941338543512682882909} = 1.5652246799676701766848237440833$$

   so $\pi^{1.5652246799676701766848237440833}$ is approximately equal to 6.   My calculator is pretty accurate.

198

The Complete Solutions and Answers   1090

1090. Each page of my history books has 3 facts, stirs 2 good emotions in me, and takes 4 calories of effort to read.

Each page of my poetry books gives me 1 fact, stirs 7 good emotions in me, and takes 2 calories of effort to read.

Tonight I want to get at least 12 facts from my reading and at least 65 good emotions. I want to spend the minimum number of calories. How should I divide my times between the history and poetry books?

Let x = the number of pages of my history books that I read.
Let y = the number of pages of my poetry books that I read.

I just had a thought. What if we turn all that English into a cute little table.

	facts	positive emotions	calories used
x pages of history	3	2	4
y pages of poetry	1	7	2

Then $3x + y \geq 12$
Then $2x + 7y \geq 65$ I like to get lots of good emotions from my reading.
I want to minimize $f(x, y) = 4x + 2y$

Son, this table thing is brilliant. It makes it a lot easier than trying to dig all the information out of the English. Please use this "cute little table" idea in solving all your future linear programming problems. I'm proud of you. Love, Mom

Thank you Mom. Love, Stan

Point-plotting: (4, 0) and (0, 12) are on $3x + y = 12$
$(0, 9\frac{2}{7})$, (32.5, 0) are on $2x + 7y = 65$

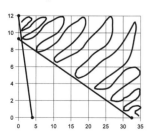

The vertices are at (0, 12), (1, 9), and (32.5, 0).
Testing them in $f(x, y) = 4x + 2y$ ...
   $f(0, 12) = 24$ calories
   $f(1, 9) = 22$ calories ⇐ the minimum
   $f(32.5, 0) = 130$ calories
      I should read one page of history and nine pages of poetry.

199

| 1095 | **The Complete Solutions and Answers**

1095. $(4\ 7\ 8\ 6)\begin{pmatrix} 9 \\ 11 \\ 8 \\ 10 \end{pmatrix} = (237)$

The cost of all the women's shoes in my Alabama store is $237.

The details: 4·9 + 7·11 + 8·8 + 6·10 = 36 + 77 + 64 + 60 = 237

Note that the columns in the left matrix are pink red black brown and they match up with the rows in the price matrix $\begin{pmatrix} 9 \\ 11 \\ 8 \\ 10 \end{pmatrix}$ ⇦ pink
⇦ red
⇦ black
⇦ brown

This will always be the case.

I, your reader, have a big question. This multiplication seems goofy. Why don't you do like you did with matrix addition. Wouldn't it be easier if you multiplied (4 7 8 6)(9 11 8 10)? I don't want to end up looking like this:

I had the same thought when I first learned matrix multiplication. It turns out that horizontal left matrix times vertical right matrix will make much more sense once you really start using matrices.

You mean that someday I'll really use this stuff?

It depends on what you do with your life. Probably none of Advanced Algebra will be of much use to you if you become a burgler, a graffiti artist, a fast-food cashier, a bum, a used-car salesman, or an illustrator of Life of Fred books.

However, if you own a business or work in the sciences, then working with matrices might come in very handy. There are computer programs that just love to add and multiply matrices. They know all about how to invert and transpose matrices and a thousand other matrix tricks. You will learn some of these tricks in *Life of Fred: Linear Algebra,* which is the book that follows calculus.

200

The Complete Solutions and Answers 1102–1122

1102. The probability (p) of seeing something educational on television varies directly as the number of hours (h) you sit in front of the television and inversely as your intelligence (i).

$$p = \frac{kh}{i} \quad \text{or it can be written} \quad p = k\frac{h}{i}$$

Question: Why does the chance of finding something educational vary inversely with intelligence?

Answer: The brighter you are, the more you probably know. The more that you know, the less likely you will encounter something educational on television.

1107.
$$(0.85)^x = \frac{1}{2}$$
$$(0.85)^x = 0.5$$

Take the log of both sides $\quad \log(0.85)^x = \log 0.5$

Birdie Rule $\quad x \log 0.85 = \log 0.5$

Divide both sides by log 0.85 $\quad x = \dfrac{\log 0.5}{\log 0.85}$

We are asked to approximate the answer to the nearest year $\quad x \approx \dfrac{-0.3010}{-0.0706} \approx 4.263 \doteq 4$ years

1112.
$\log_{44} 44 = 1 \quad$ since $44^1 = 44$
$\log_8 64 = 2 \quad$ since $8^2 = 64$
$\log_{1/2} 2 = -1 \quad$ since $(\tfrac{1}{2})^{-1} = 2$

1117. $\dfrac{\log_{11} 16}{\log_{11} 0.25}$

$= \log_{1/4} 16 \quad$ using $\dfrac{\log_c a}{\log_c b} = \log_b a$

$= -2 \quad$ since $(\tfrac{1}{4})^{-2} = 4^2 = 16$

1122. $\dfrac{44 + 24i}{4} = \dfrac{44}{4} + \dfrac{24i}{4} = 11 + 6i$

| 1127–1130 | The Complete Solutions and Answers

1127. What is the distance between (–3, 5) and (3, –5)?

$$d = \sqrt{(3-(-3))^2 + (-5-5)^2} = \sqrt{36 + 100}$$
$$= \sqrt{136}$$
$$= \sqrt{4}\sqrt{34} \quad \text{Simplifying the square root was}$$
$$= 2\sqrt{34} \quad \text{optional.}$$

1130. Find the sum of each of these:

$$\sum_{i=1}^{44} 6 + i = (6+1) + (6+2) + (6+3) + \ldots + (6+44)$$

$$= 7 + 8 + 9 + \ldots + 50$$

An arithmetic series with a = 7, d = 1, ℓ = 50, n = 44.

$$s = \frac{n}{2}(a + \ell) = 22(7 + 50) = 1{,}254$$

$$\sum_{i=3}^{72} 4i = 12 + 16 + 20 + \ldots + (4)(72)$$

An arithmetic series with a = 12, d = 4, ℓ = (4)(72), n = 70

Careful! It is not n = 69.

Look at the first series. $\sum_{i=1}^{44}$ There are 44 terms. We didn't subtract 44 – 1.

$$s = \frac{n}{2}(a + \ell) = 35(12 + 288) = 10{,}500$$

$$\sum_{i=1}^{\infty} \left(\frac{3}{8}\right)^i = \frac{3}{8} + \left(\frac{3}{8}\right)^2 + \left(\frac{3}{8}\right)^3 + \ldots$$

An infinite geometric series with a = $\frac{3}{8}$, r = $\frac{3}{8}$

We first make sure that $-1 < \frac{3}{8} < 1$.

Fancy algebra books write $|r| < 1$ instead of $-1 < r < 1$.

$$s = \frac{a}{1-r} = \frac{3}{8} \div \frac{5}{8} = \frac{3}{8} \times \frac{8}{5} = \frac{3}{5}$$

The Complete Solutions and Answers 1137–1142

1137. Graph these six lines.

$y = \frac{2}{3}x + 4 \quad m = \frac{2}{3} \quad \text{y-intercept} = (0, 4)$

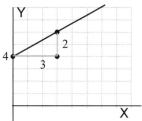

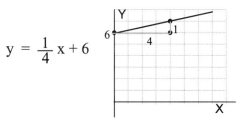

$y = \frac{1}{4}x + 6$

$y = \frac{-5}{3}x + 2$

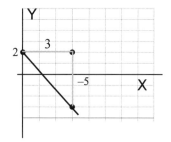

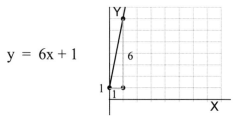

$y = 6x + 1$

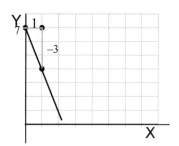

$y = -3x + 7$

$y = x + 5$

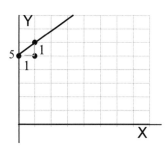

1142. Factor $7x^2 + 21x + 14 = 7(x^2 + 3x + 2) = 7(x + 2)(x + 1)$

Always look for a common factor first.

$y^2 - 2y - 24 = (y + 4)(y - 6)$
$x^2 + 27xy + 50y^2 = (x + 25y)(x + 2y)$
$w^2 + 5w - 50 = (w + 10)(w - 5)$

1147–1162 The Complete Solutions and Answers

1147. If (a, b) lies in quadrant I, what can you say about a + b?
Answer: Points in quadrant I have positive x-coordinates and positive y-coordinates, so a + b > 0.

If (a, b) lies in quadrant II, want can you say about ab?
Answer: If (a, b) is in quadrant II, then a < 0 and b > 0, so ab < 0.

Give an example of a point with a positive ordinate that does not lie in either quadrant I or quadrant II.
Answer: Any point on the positive Y axis, such as (0, 6) or (0, 3.98), has a positive ordinate (y-coordinate) but does not lie in either quadrant I or quadrant II.

1152. What is the equation of the line with a slope of $\sqrt{2}$ that passes through the point (33, 55)?

$$m = \frac{y - y_1}{x - x_1} \quad \text{becomes} \quad \sqrt{2} = \frac{y - 55}{x - 33}$$

1157. Solve $\begin{cases} 3x - 5y = 7 \\ 2x - y = 7 \end{cases}$

Solve the second equation for y $\qquad 2x - 7 = y$
Substitute y = 2x − 7 into the first equation $\qquad 3x - 5(2x - 7) = 7$

$$3x - 10x + 35 = 7$$
$$28 = 7x$$
$$4 = x$$

Substitute x = 4 into 2x − 7 = y $\qquad 2(4) - 7 = y$
$$1 = y$$

1162. What are the coordinates of the center and of the vertices of the hyperbola $-\frac{(x-7)^2}{100} + \frac{(y+23)^2}{144} = 1$

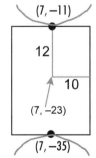

The center is still (7, −23).
The egg box still has the same shape.

The only difference is that the hyperbola extends up and down (in the y direction) rather than left and right.
The vertices of the hyperbola are at the top and bottom of the box.
The vertices are located at (7, −23 ± 12)

The Complete Solutions and Answers

1167

1167. Each page of my history books has 3 facts, stirs 2 good emotions in me, and take 4 calories of effort to read. Each page of my poetry books gives me 1 fact, stirs 7 good emotions in me, and takes 2 calories of effort to read.

I have just eaten a tiny piece of chocolate and have 20 calories to spend on reading. I still want to get at least 12 facts from my reading.

How should I divide my times between the history and poetry books in order to maximize my good emotions?

We can use the same cute little table.

	facts	positive emotions	calories used
x pages of history	3	2	4
y pages of poetry	1	7	2

I have only 20 calories to spend on reading $4x + 2y \leq 20$
I want at least 12 facts from my reading $3x + y \geq 12$
I want to maximize $f(x, y) = 2x + 7y$

Point-Plotting (5, 0) and (0, 10) are on $4x + 2y = 20$
        (4, 0) and (0, 12) are on $3x + y = 12$

$4x + 2y \leq 20$ will be shaded to the lower left. ⎱ I know these facts by testing points
$3x + y \geq 12$ will be shaded to the upper right. ⎰ on each side of the lines.

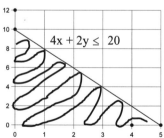

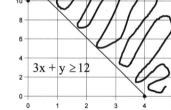

Testing the vertices in $f(x, y) = 2x + 7y$

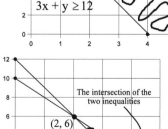

$f(2, 6) = 46$ good emotions
$f(4, 0) = 8$ good emotions
$f(5, 0) = 10$ good emotions

The intersection of the two inequalities

(2, 6)

I should read 2 pages of history and 6 pages of poetry.

1172–1177 The Complete Solutions and Answers

1172. Solve $\dfrac{2x}{4-x} = \dfrac{3+x}{5x}$

      Cross multiply $10x^2 = (4-x)(3+x)$
      Multiply $10x^2 = 12 + 4x - 3x - x^2$

This is a quadratic equation. The easiest way to solve quadratic equations is by factoring—if it factors.

      Set everything equal to zero $11x^2 - x - 12 = 0$

This is a trinomial of the form $ax^2 + bx + c$ where $a \neq 1$. Using the "quicker, handy-dandy approach" (from *Life of Fred: Beginning Algebra Expanded Edition,* page 307), we split $-x$ into two numbers that add to $-x$ and that multiply to $(11x^2)(-12)$, which is $-132x^2$.

If you know your multiplication tables up to 12 x 12, the number 132 has one obvious factoring.

      $11x$ and $-12x$ multiply to $-132x^2$ and add to $-x$.
      Split $-x$ into $11x$ and $-12x$ $11x^2 + 11x - 12x - 12 = 0$
      Factor by grouping $11x(x+1) - 12(x+1) = 0$
                                       $(x+1)(11x-12) = 0$
      Set each factor equal to zero $x + 1 = 0$ OR $11x - 12 = 0$
      Solve $x = -1$ OR $x = 12/11$

1177. If you look at a circle from an angle, it can become an ellipse.

But the sun is a sphere (a ball). Look at it from any direction and it will always appear as a circle.

Let's play for a moment.
If you draw an ellipse, can you look at it from an angle and make it look like a circle?

    Answer: If you look at the ellipse from over here, you can make it look like a circle.

(Continued on next page)

206

The Complete Solutions and Answers

Second question: If you draw two parallel lines, is it possible to look at them from an angle so that they do not look parallel?
Answer: Yes. The lines in this hallway are parallel, but if you stand in the hallway and look, they appear to intersect at a point just beyond the people at the end of the hall.

One way to think of "looking at something from an angle" is to compare the actual object with a photo of that object.

What things stay true in the photo?
Answer: Certainly being a circle is not *a projective invariant*. (**Projective invariant** is the official math language.)

Certainly having right angles is not a projective invariant. Here is a picture that I just took of the cover of *Life of Fred: Geometry*. In this picture, none of the corners are right angles.

Here are some things that are projective invariants:
❋ Points (A picture of a point is a point.)
❋ Lines
❋ Betweenness (If we have a segment AB and point C is somewhere between A and B, then C will be between A and B in the photo.)

There are many different kinds of geometries. The main one studied in high school geometry is called Euclidean geometry.

The one that talks about projective invariants is called Projective geometry. Painters, photographers, graphic designers, and many others think that Projective geometry is "more true" than Euclidean geometry.

Here are four basic assumptions for Projective geometry:
1. If P and Q are two distinct points, then there is exactly one line that passes through them. (This assumption is also true in Euclidean geometry.)
2. There are at least three points on any line. (Also true for Euclidean geometry.)
3. Not all points are on the same line. (Also true for Euclidean geometry.)
4. If ℓ and m are distinct lines, then there is at least one point on both of them! (No parallel lines.)

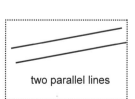
two parallel lines

207

| 1180 | The Complete Solutions and Answers

1180. On a day I go barefoot, my arches improve by 3 points, I get 8 ouches, and make 5 new friends.

On a day I wear shoes, my arches improve by 1 point, I get 2 ouches, and make 4 new friends.

I want my arches to improve by at least 18 points and I want to get no more than 24 ouches. How many days should I go barefoot and how many days should I wear shoes in order to make as many new friends as possible?

My cute little table . . .

	arch improvement	ouches	new friends
x days of barefoot	3	8	5
y days of shoes	1	2	4

I want my arches to improve by at least 18 points $3x + y \geq 18$
I want no more than 24 ouches $8x + 2y \leq 24$
I want to maximize $f(x, y) = 5x + 4y$

Graphing $3x + y \geq 18$ Graphing $8x + 2y \leq 24$

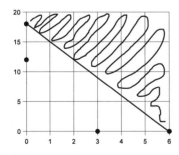 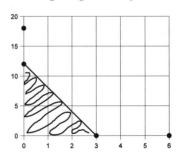

So I want lots of arch improvement and very few ouches. I am not going to get it. Those two graphs have no overlap. They do not intersect.

No matter how many days I walk barefoot and how many days I walk with shoes on, I can never achieve both 18 points of arch improvement and have less than 24 ouches.

208

The Complete Solutions and Answers

1187. There are 300 flavors of ice cream. A survey was taken to determine the 100 most popular flavors. The results of the survey declared the first place winner, the second place winner, and so on, all the way down to the 100th most popular flavor.

How many ways could this survey have turned out?

There are 300 choices for the first place winner.
There are 299 choices for the second place winner.
There are 298 choices for the third place winner.
. . .
There are 201 choices for the 100th place winner.

There are 300×299×298× . . . ×202×201 possible ways the survey could turn out.

$$300 \times 299 \times 298 \times \ldots \times 202 \times 201 = \frac{300!}{200!}$$

I hope this is starting to look familiar. This is exactly the same as the previous problem: *How many possible functions are there for g if g:A → B where A has 100 members and B has 300 members and g is 1-1?*

In both problems, the answer is P(300, 100), which is $\frac{300!}{200!}$

In general, $P(n, r) = \frac{n!}{(n-r)!}$ If I have a population of n items and I am going to select r of them *in a particular order*, then there are P(n, r) ways to do that.

If there are 300 cities and I want to plan a vacation to 100 of them in a particular order, then there are P(300, 100) ways to plan that vacation.

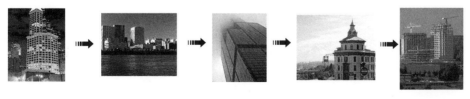

etc.

1192–1207 | The Complete Solutions and Answers

1192. $(3i)(4)(-5i)(2)$
$= (12i)(-5i)(2)$
$= (-60i^2)(2)$
$= (60)(2)$
$= 120$

1197. Draw a Venn diagram of the intersection of all log cabins (L) and the set of all buildings (B).

The intersection of all log cabins and all buildings are those things that are both log cabins and buildings. Since every log cabin is a building, $L \cap B = L$.

1202. Evaluate $\begin{vmatrix} 1 & 0 & 30 \\ 5 & 9 & -3 \\ 4 & 2 & -8 \end{vmatrix}$

I choose column 2.

$$-(0)\begin{vmatrix} 5 & -3 \\ 4 & -8 \end{vmatrix} + (9)\begin{vmatrix} 1 & 30 \\ 4 & -8 \end{vmatrix} - (2)\begin{vmatrix} 1 & 30 \\ 5 & -3 \end{vmatrix}$$

$= 0 + 9(-8 - 120) - 2(-3 - 150)$
$= 9(-128) - 2(-153)$
$= -846$

1207. Complete the square for each of these.

$x^2 + 6x$ Half of +6 is +3. Three squared is 9. $x^2 + 6x + 9$ which factors into $(x + 3)^2$

$y^2 - 10y$ Half of −10 is −5. Minus 5 squared is +25 $y^2 - 10y + 25$ which equals $(y - 5)^2$

$5z^2 + 40z$ There must be a 1 in front of the z^2 in order to complete the square. Factor out 5 first.
$\qquad 5z^2 + 40z = 5(z^2 + 8z)$ Then complete the square.
$\qquad\qquad 5(z^2 + 8z + 16)$
$\qquad\qquad 5(z + 4)^2$

$9w^2 + 36w = 9(w^2 + 4w)$
$\qquad\qquad 9(w^2 + 4w + 4)$
$\qquad\qquad 9(w + 2)^2$

The Complete Solutions and Answers

1212. Resolve into partial fractions $\dfrac{7x^2 + 6x - 21}{(x-3)(x+2)(x+1)}$

$$\dfrac{7x^2 + 6x - 21}{(x-3)(x+2)(x+1)} = \dfrac{A}{x-3} + \dfrac{B}{x+2} + \dfrac{C}{x+1}$$

Multiply each term by $(x-3)(x+2)(x+1)$

$$7x^2 + 6x - 21 = A(x+2)(x+1) + B(x-3)(x+1) + C(x-3)(x+2)$$

Let $x = 3$
$$7(3)^2 + 6(3) - 21 = A(3+2)(3+1)$$
$$60 = 20A$$
$$3 = A$$

Let $x = -2$
$$7(-2)^2 + 6(-2) - 21 = B(-2-3)(-2+1)$$
$$-5 = 5B$$
$$-1 = B$$

Let $x = -1$
$$7(-1)^2 + 6(-1) - 21 = C(-1-3)(-1+2)$$
$$-20 = -4C$$
$$5 = C$$

$$\dfrac{7x^2 + 6x - 21}{(x-3)(x+2)(x+1)} = \dfrac{3}{x-3} + \dfrac{-1}{x+2} + \dfrac{5}{x+1}$$

1217. $\begin{pmatrix} 3 & 9 \\ 5 & 2 \\ -2 & 4 \end{pmatrix} \begin{pmatrix} 3 & 2 & 4 & 9 \\ 9 & 1 & 1 & 7 \end{pmatrix}$

$= \begin{pmatrix} (3)(3)+(9)(9) & (3)(2)+(9)(1) & (3)(4)+(9)(1) & (3)(9)+(9)(7) \\ (5)(3)+(2)(9) & (5)(2)+(2)(1) & (5)(4)+(2)(1) & (5)(9)+(2)(7) \\ (-2)(3)+(4)(9) & (-2)(2)+(4)(1) & (-2)(4)+(4)(1) & (-2)(9)+(4)(7) \end{pmatrix}$

$= \begin{pmatrix} 90 & 15 & 21 & 90 \\ 33 & 12 & 22 & 59 \\ 30 & 0 & -4 & 10 \end{pmatrix}$

> 1222–1225

The Complete Solutions and Answers

1222. Solve $\frac{2}{x-8} = \frac{x-1}{-3}$

   Cross multiply $-6 = (x-8)(x-1)$
   Multiply $-6 = x^2 - x - 8x + 8$

This is a quadratic equation. The easiest way to solve quadratic equations is by factoring—if it factors.

   Set everything equal to zero $0 = x^2 - 9x + 14$

Find two numbers that add to –9 and multiply to 14 $0 = (x-2)(x-7)$

   Set each factor equal to zero $x - 2 = 0$ OR $x - 7 = 0$
   Solve $x = 2$ OR $x = 7$

1225. How high (h) a fireworks rocket goes varies directly as the square root of the amount of powder (p) in the rocket.

  A rocket with 38 pounds of powder will ascend 100 feet. How high will a rocket with 57 pounds of powder ascend?

Step ①: Find the equation. $h = k\sqrt{p}$
Step ②: Find the value of k.
If p = 38, then h = 100.
Substituting that into $h = k\sqrt{p}$, we get ☞ $100 = k\sqrt{38}$

Divide both sides by $\sqrt{38}$ $\frac{100}{\sqrt{38}} = k$

$h = k\sqrt{p}$ now becomes $h = \frac{100}{\sqrt{38}}\sqrt{p}$

Step ③: Find h when p = 57 $h = \frac{100\sqrt{57}}{\sqrt{38}}$

For fun, if we want to rationalize the denominator (which we did on page 406 of Life of Fred: Beginning Algebra Expanded Edition), we multiply the top and bottom by $\sqrt{38}$ and obtain

$$\frac{100\sqrt{57}\,\sqrt{38}}{\sqrt{38}\,\sqrt{38}}$$

   which equals $\frac{100\sqrt{57}\,\sqrt{38}}{38}$

   or, if you like, $\frac{100\sqrt{2166}}{38}$ or $\frac{50\sqrt{2166}}{19}$

The Complete Solutions and Answers | 1228

1228. On a day I go barefoot, my arches improve by 3 points, I get 8 ouches, and make 5 new friends. On a day I wear shoes my arches, improve by 1 point, I get 2 ouches, and make 4 new friends.

I want my arches to improve by at least 18 points and I want to make at least 40 new friends.

How many days should I go barefoot and how many days should I wear shoes in order to minimize my ouches?

We have the same cute little table . . .

	arch improvement	ouches	new friends
x days of barefoot	3	8	5
y days of shoes	1	2	4

These are called the constraints:

I want my arches to improve by at least 18 points. $3x + y \geq 18$
I want to make at least 40 new friends. $5x + 4y \geq 40$

This is called the objective function:

I want to minimize my ouches. $f(x, y) = 8x + 2y$

Point-plotting: $(0, 18)$ and $(6, 0)$ are on $3x + y = 18$.
$(0, 10)$ and $(8, 0)$ are on $5x + 4y = 40$.

Using the elimination method, $(4\frac{4}{7}, 4\frac{2}{7})$ satisfies both equations.

Plotting $x \geq 0$, $y \geq 0$, $3x + y \geq 18$, and $5x + 4y \geq 40$

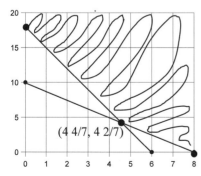

Testing the vertices in the objective function $f(x, y) = 8x + 2y$

$f(0, 18) = 36$ ⇐ the minimum
$f(4\frac{4}{7}, 4\frac{2}{7}) = 45\frac{1}{7}$
$f(8, 0) = 64$

To improve my arches by at least 18 points and to make at least 40 new friends, I will have the minimum number of ouches if I go barefoot zero days and wear shoes 18 days.

| 1234–1240 | The Complete Solutions and Answers

1234. Find the mean and median averages of 300, 700, 500.

To find the mean average of three numbers, add them up and divide the sum by three. $300 + 700 + 500 = 1500$ $1500 \div 3 = 500$

To find the median average, sort the numbers from smallest to largest and pick the one in the middle.   300, 500, 700

The median average is 500.

1237. Evaluate $\begin{vmatrix} 6 & 2 & 7 & 1 \\ 3 & 0 & 0 & 6 \\ 9 & 1 & 0 & 4 \\ 9 & 2 & 0 & 3 \end{vmatrix}$

Who could resist column 3? It has three zeros in it.

$+7 \begin{vmatrix} 3 & 0 & 6 \\ 9 & 1 & 4 \\ 9 & 2 & 3 \end{vmatrix} - 0 \begin{vmatrix} 6 & 2 & 1 \\ 9 & 1 & 4 \\ 9 & 2 & 3 \end{vmatrix} + 0 \text{ (its minor)} \quad -0 \text{ (its minor)}$

Wait a minute. It's silly to write ⟶ It is going to be multiplied by zero.

$= \quad +7 \begin{vmatrix} 3 & 0 & 6 \\ 9 & 1 & 4 \\ 9 & 2 & 3 \end{vmatrix}$

Expanding by the first row

$= \quad 7 \left(+3 \begin{vmatrix} 1 & 4 \\ 2 & 3 \end{vmatrix} - 0 \begin{vmatrix} 9 & 4 \\ 9 & 3 \end{vmatrix} + 6 \begin{vmatrix} 9 & 1 \\ 9 & 2 \end{vmatrix} \right)$

$= \quad 7 \left(3(3-8) \qquad\qquad + 6(18-9) \right)$

$= \quad 7(-15 + 54) \quad = \quad 273$

1240. The set of all elements in the codomain that are the image of at least one element in the domain is called the __range__ of the function.

Combining all the words together: The range of a function is always a subset of the codomain. If the range of a function is equal to the codomain, then the function is onto its codomain.

The Complete Solutions and Answers |1246–1258|

1246. $\sqrt{18} = \sqrt{9}\sqrt{2} = 3\sqrt{2}$

$\sqrt{x^5 y} = \sqrt{x^4}\sqrt{xy} = x^2\sqrt{xy}$

$1000^{1/3} = 10$

1249. $\dfrac{x-y}{\sqrt{x-y}} = \dfrac{x-y}{\sqrt{x-y}} \cdot \dfrac{\sqrt{x-y}}{\sqrt{x-y}} = \dfrac{(x-y)\sqrt{x-y}}{x-y} = \sqrt{x-y}$

1252. $(\log_8 9)(\log_9 64) = \log_8 64$ using $(\log_c b)(\log_b a) = \log_c a$
                                 $= 2$ since $8^2 = 64$

1255. Place into standard form for an ellipse $5x^2 + y^2 + 6y = 16$

Add $+9$ to both sides of the equation in order to complete the square $5x^2 + y^2 + 6y + 9 = 25$

Factor $y^2 + 6y + 9$ $5x^2 + (y+3)^2 = 25$

Divide each term by 25 (in order to put a 1 on the right side of the equation) $\dfrac{x^2}{5} + \dfrac{(y+3)^2}{25} = 1$

Put into standard form $\dfrac{x^2}{(\sqrt{5})^2} + \dfrac{(y-(-3))^2}{5^2} = 1$

This is an ellipse with center at $(0, -3)$.
The semi-major axis is vertical and has a length of 5.
The semi-minor axis is horizontal and has a length of $\sqrt{5}$.

    If you were to graph it, you would need to graph the approximate value of $\sqrt{5}$ which is 2.2360679774997896964091736687313. Most people's eyes and hands are incapable of this much accuracy. They usually graph $\sqrt{5} \doteq 2.2$ instead.

    English lesson. Originally, I wrote, "They usually graph $\sqrt{5} \doteq 2.2$." The first dot is a decimal point and the second dot is a period. But "2.2." looked funny even though it is probably correct. I fluffed out the sentence by adding "instead."

    My English teacher told me that in addition to writing sentences that are correct, it is nice not to drive your readers crazy.

1258. On the first day of class, the music teacher told the students to practice their violins for a certain number of minutes that night. On the next day of class, he told them to practice for 4 minutes more. On the third day of class he told them to practice for 4 minutes more than the previous night. Each night they practiced 4 minutes longer than the previous night. On the 66th day of class, they were told to practice for 277 minutes that night. How long did they practice on the first night?

It's an arithmetic sequence. $\ell = a + (n-1)d$ becomes $277 = a + (65)4$
        They practiced for 17 minutes $277 = a + 260$
        on the first night. $17 = a$

1261–1264 The Complete Solutions and Answers

1261. Which of these can be solved by cross multiplication?

First example: $\dfrac{x^2+6}{5x} = \dfrac{3.388x}{921}$ ☒ yes ☐ no

Second example: $\dfrac{3}{7x^2} = \dfrac{5}{x-2} = \dfrac{x+6}{8}$ ☐ yes ☒ no

Cross multiplying is done with one equation, not two.

Third example: $\dfrac{\pi x}{7} = \dfrac{4}{9}$ ☒ yes ☐ no

Pi (π) is a number just like 7 or 398. The solution to this third example is easy.

Cross multiply $\qquad 9\pi x = 28$
Divide both side by $9\pi \qquad x = \dfrac{28}{9\pi}$

Fourth example: $\dfrac{5+x}{3} = \dfrac{7x}{9} + 4$ ☐ yes ☒ no

(unless you first add the 7x/9 and the 4)

Cross multiplying is done when you have a proportion.
Translation: Cross multiplying is done when you have two ratios equal to each other.
Translation: Cross multiplying is done when you have two fractions equal to each other: $\dfrac{a}{b} = \dfrac{c}{d}$

1264.

$\begin{pmatrix} 3 & 5 & 11 & 2 \\ 4 & 7 & 8 & 6 \\ 1 & 0 & 23 & 7 \end{pmatrix} \begin{pmatrix} 9 \\ 11 \\ 8 \\ 10 \end{pmatrix} = \begin{pmatrix} 190 & 237 & 263 \end{pmatrix}$

pink red black brown

The details: $3\cdot 9 + 5\cdot 11 + 11\cdot 8 + 2\cdot 10 = 190$
$1\cdot 9 + 0\cdot 11 + 23\cdot 8 + 7\cdot 10 = 263$

Multiplying the **first** row times the **first** column gives the entry in the **first** row, **first** column of the answer.

Multiplying the **second** row times the **first** column gives the entry in the **second** row, **first** column of the answer.

Multiplying the **third** row times the **first** column gives the entry in the **third** row, **first** column of the answer.

In general, multiplying the 5th row times the 6th column gives the entry in the 5th row, 6th column of the answer.

Fancy algebra books would say: In general, multiplying the ith row times the kth column gives the entry in the ith row, k^{th} column of the answer.

The Complete Solutions and Answers 1270–1276

1270. $\sqrt{x + 7x^2 + 3823} = -17$

There is no solution. A $\sqrt{}$ can never equal a negative number. (See principal square roots in Life of Fred: Beginning Algebra Expanded Edition, page 380.)

Don't you just love those one-step problems!

1273. Solve $6 = \sqrt{13x - x^2}$

$36 = 13x - x^2$	Square both sides
$x^2 - 13x + 36 = 0$	Solve by factoring
$(x - 4)(x - 9) = 0$	Two numbers that add to –13 and multiply to +36

$x - 4 = 0$ OR $x - 9 = 0$
$x = 4$ OR $x = 9$

When you square both sides of an equation, you must check your answers in the original equation.

Checking x = 4
$6 \stackrel{?}{=} \sqrt{13(4) - 16}$
$6 \stackrel{?}{=} \sqrt{36}$ yes

Checking x = 9
$6 \stackrel{?}{=} \sqrt{13(9) - 81}$
$6 \stackrel{?}{=} \sqrt{117 - 81}$
$6 \stackrel{?}{=} \sqrt{36}$ yes

The solution to $6 = \sqrt{13x - x^2}$ is $x = 4$ and $x = 9$.

1276. Graph $\dfrac{2}{5} = \dfrac{y - 6}{x - 4}$

This is a line through (4, 6) with a slope of $\dfrac{2}{5}$

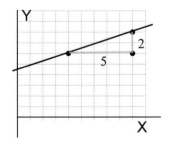

The Complete Solutions and Answers

1279. $\dfrac{8}{\sqrt{w}} = \dfrac{8}{\sqrt{w}} \cdot \dfrac{\sqrt{w}}{\sqrt{w}} = \dfrac{8\sqrt{w}}{w}$

In the old days before hand-held calculators, rationalizing the denominator was very handy. If you had an expression like $\dfrac{5}{\sqrt{3}}$ you faced a division problem like $\dfrac{5}{1.73205080756887729}$

which looked like $1.73205080756887729 \overline{)5.0000000000000000000}$ ↩ miserable!

Instead, if you rationalized the denominator $\dfrac{5}{\sqrt{3}}$ would turn into $\dfrac{5\sqrt{3}}{3}$ (after you multiplied top and bottom by $\sqrt{3}$).

Approximating $\dfrac{5\sqrt{3}}{3}$ by $\dfrac{(5)(1.73205080756887729)}{3}$ is something you can do in a jiffy.

1282. $(-1 + \sqrt{3}\, i)^3$ This has a surprising answer.

$$
\begin{aligned}
(-1 + \sqrt{3}\,i)^3 &= (-1 + \sqrt{3}\,i)(-1 + \sqrt{3}\,i)(-1 + \sqrt{3}\,i) \\
&= (1 - \sqrt{3}\,i - \sqrt{3}\,i - 3)(-1 + \sqrt{3}\,i) \\
&= (-2 - 2\sqrt{3}\,i)(-1 + \sqrt{3}\,i) \\
&= (2 - 2\sqrt{3}\,i + 2\sqrt{3}\,i + 6) \\
&= 8
\end{aligned}
$$

That means that one of the cube roots of 8 is $(-1 + \sqrt{3}\,i)$.
Another cube root of 8 is $(-1 - \sqrt{3}\,i)$.
And the cube root of 8 that you already know is 2.

There are three different numbers, which when cubed, equal 8.

> Fact: There are 60 different numbers, which when raised to the 60th power, equal 5.
> That means that $x^{60} = 5$ has 60 different solutions.

Your question: Why didn't anyone tell me about this?
My answer: Fifty-nine of the 60 solutions to $x^{60} = 5$ will involve i. You just learned about i.

Your question: How do I find the other 59 solutions of $x^{60} = 5$?
My answer: To find those other 59 solutions of $x^{60} = 5$ you need more than just i. You need the sine and the cosine functions that are in trig. In Chapter 10 of Life of Fred: Trig Expanded Edition, in one of the Your Turn to Play sections, you will find the million answers to $x^{1,000,000} = i$. In English, the million millionth roots of i.

1285. Draw the Venn diagram of the set of all humans (H) and the set of all people who currently live in Paris (P).

Since every person who currently lives in Paris is a human, we know that P is a subset of H.

$P \subset H.$

The Complete Solutions and Answers | 1288–1291

1288. Place into standard form $\quad 4x^2 - 16x + 9y^2 - 90y = -205$

We need ones in front of x^2 and y^2 $\quad 4(x^2 - 4x) + 9(y^2 - 10y) = -205$

Complete the square $\quad 4(x^2 - 4x + 4) + 9(y^2 - 10y + 25) = -205 + 16 + 225$

When we added +4 inside the parentheses, we were adding 16 to the left side of the equation.
When we added +25 inside the parentheses, we were adding 225 to the left side of the equation.

Factor $\quad 4(x - 2)^2 + 9(y - 5)^2 = 36$

Divide each term by 36 in order to put a 1 on the right side of the equation $\quad \dfrac{(x-2)^2}{9} + \dfrac{(y-5)^2}{4} = 1$

into standard form $\quad \dfrac{(x-2)^2}{3^2} + \dfrac{(y-5)^2}{2^2} = 1$

1291. What is the equation of the line that is perpendicular to the line through (2, 5) and (4, 8) and that passes through (8, 9)?

This is a three-step problem.

First, find the slope of the line that passes through (2, 5) and (4, 8).

> The slope formula given two points (x_1, y_1) and (x_2, y_2)
> $$m = \dfrac{y_2 - y_1}{x_2 - x_1}$$

$$m = \dfrac{8-5}{4-2} = \dfrac{3}{2}$$

Second, find the slope of the line perpendicular to a line with slope $\dfrac{3}{2}$

Two lines are perpendicular if their slopes are the negative reciprocals of each other. $\dfrac{a}{b}$ and $\dfrac{-b}{a}$

The negative reciprocal of $\dfrac{3}{2}$ is $\dfrac{-2}{3}$

Third, given a point (8, 9) and a slope $\dfrac{-2}{3}$ we use the point-slope formula

$$m = \dfrac{y - y_1}{x - x_1} \quad \text{becomes} \quad \dfrac{-2}{3} = \dfrac{y - 9}{x - 8}$$

or $\quad 2x + 3y = 43$

1294–1297 The Complete Solutions and Answers

1294. Find the center and vertices of $100x^2 + 800x + 9y^2 + 90y = -925$

In order to complete the square
the coefficients of x^2 and y^2 must be 1. $100(x^2 + 8x) + 9(y^2 + 10y) = -925$

Adding 16 inside the parentheses is
the same as adding 1600 to the left
side of the equation.

$$100(x^2 + 8x + 16) + 9(y^2 + 10y + 25) = -925 + 1600 + 225$$

$$100(x + 4)^2 + 9(y + 5)^2 = 900$$

Divide both sides by 900 $\dfrac{(x + 4)^2}{9} + \dfrac{(y + 5)^2}{100} = 1$

The center of this ellipse is at $(-4, -5)$.
The major axis is vertical
(in the y direction) since $100 > 9$.
The vertices are at $(-4, -5 \pm 10)$.

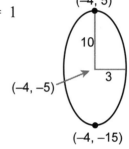

1297. Find the sum of each of these.

$$\sum_{i=1}^{8} (0.3)^i = 0.3 + (0.3)^2 + (0.3)^3 + \ldots + (0.3)^8$$

A geometric series with $a = 0.3$, $r = 0.3$, and $n = 8$

$$s = \dfrac{a(1 - r^n)}{1 - r} = \dfrac{0.3(1 - 0.3^8)}{0.7} = \dfrac{0.3(1 - 0.00006561)}{0.7}$$

$$= \dfrac{0.299980317}{0.7} \approx 0.42854331$$

$$\sum_{i=1}^{\infty} 5\left(\dfrac{1}{4}\right)^i = \dfrac{5}{4} + \dfrac{5}{16} + \dfrac{5}{64} + \ldots$$

An infinite geometric series with $a = \dfrac{5}{4}$ and $r = \dfrac{1}{4}$

$$s = \dfrac{a}{1 - r} = \dfrac{5}{4} \div \dfrac{3}{4} = \dfrac{5}{4} \times \dfrac{4}{3} = \dfrac{5}{3} \text{ or } 1\dfrac{2}{3}$$

$$\sum_{i=5}^{55} (\pi + i) = (\pi + 5) + (\pi + 6) + \ldots + (\pi + 55)$$

An arithmetic series with $a = \pi + 5$, $d = 1$, and $n = 51$.

$$s = \dfrac{n}{2}(a + \ell) = \dfrac{51}{2}(\pi + 5 + \pi + 55) = 51\pi + 1{,}530$$

The Complete Solutions and Answers | 1300–1309

1300. Solve $\dfrac{3x+1}{2x} = x$

Using the hint given in the problem $\quad \dfrac{3x+1}{2x} = \dfrac{x}{1}$

Cross multiply $\quad\quad\quad\quad\quad\quad\quad\quad 3x + 1 = 2x^2$

This is a quadratic equation. The easiest way to solve quadratic equations is by factoring—if it factors.

Set everything equal to zero $\quad\quad\quad 0 = 2x^2 - 3x - 1$

It doesn't factor. ☹ The quadratic formula will always work.

$ax^2 + bx + c = 0$ has the solution $x = \dfrac{-b \pm \sqrt{b^2 - 4ac}}{2a}$

Use the quadratic formula $\quad\quad\quad x = \dfrac{3 \pm \sqrt{9 - (4)(2)(-1)}}{4}$

Do the arithmetic $\quad\quad\quad\quad\quad\quad x = \dfrac{3 \pm \sqrt{17}}{4}$

1303. $\dfrac{x-y}{\sqrt{x} - \sqrt{y}} = \dfrac{x-y}{\sqrt{x} - \sqrt{y}} \cdot \dfrac{\sqrt{x} + \sqrt{y}}{\sqrt{x} + \sqrt{y}}$

$\quad\quad\quad\quad = \dfrac{(x-y)(\sqrt{x} + \sqrt{y})}{x - y}$

$\quad\quad\quad\quad = \sqrt{x} + \sqrt{y}$

1306. Factor $\quad 18w^3 + 27w^4 = 9w^3(2 + 3w)$
$\quad\quad\quad\quad\quad 3x^9 - 12x^6y = 3x^6(x^3 - 4y)$
$\quad\quad\quad\quad\quad 15xyz + 20x = 5x(3yz + 4)$

1309. If I buy 3 dogs and 2 bags of dog food, it will cost $70.
If I buy 4 dogs and 3 bags of dog food, it will cost $97.

$\quad\quad\quad \begin{cases} 3x + 2y = 70 \\ 4x + 3y = 97 \end{cases}$

Multiply the first equation by 3. $\quad\quad\quad \begin{cases} 9x + 6y = 210 \\ -8x - 6y = -194 \end{cases}$
Multiply the second equation by –2.

Add the equations. $\quad\quad\quad\quad\quad\quad\quad\quad\quad x = 16$

Back substitute $x = 16$ into any
equation containing x and y.
I choose the first equation. $\quad\quad 3(16) + 2y = 70$
$\quad\quad\quad\quad\quad\quad\quad\quad\quad\quad\quad 2y = 22$
$\quad\quad\quad\quad\quad\quad\quad\quad\quad\quad\quad\; y = 11$

One dog cost $16 and one bag of dog food cost $11.
(Sometimes real dogs cost more than $16. I wanted to keep the numbers easy.)

1313–1318 The Complete Solutions and Answers

1313. Graph $y = 3x^2 - 5$ from $x = -2$ to $x = 2$.

x	y
-2	7
-1	-2
0	-5
1	-2
2	7

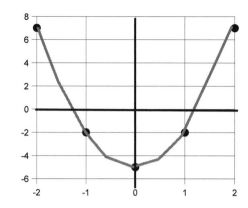

1315. Let the domain of function g be all real numbers. Define g by the rule $g(x) = \sqrt{x^2 + 2}$. Show that g is *not* 1-1.

To show that g is not 1-1, we need to find two members of the domain that have the same image in the codomain.

The x^2 is the key to this.

$g(692)$ will equal $g(-692)$.

Or, for a simpler example, $g(1) = g(-1)$.

1318. $\begin{pmatrix} 6 & 0 & -2 \\ 2 & 1 & 16 \end{pmatrix} \begin{pmatrix} 3 & 5 & 2 \\ 4 & 1 & 0 \\ 2 & 0 & 1 \end{pmatrix}$

$= \begin{pmatrix} (6)(3) + 0(4) + (-2)(2) & (6)(5) + (0)(1) + (-2)(0) & (6)(2) + (0)(0) + (-2)(1) \\ (2)(3) + (1)(4) + (16)(2) & (2)(5) + (1)(1) + (16)(0) & (2)(2) + (1)(0) + (16)(1) \end{pmatrix}$

$= \begin{pmatrix} 14 & 30 & 10 \\ 42 & 11 & 20 \end{pmatrix}$

The Complete Solutions and Answers

1321. Ice cream! The menu suggested either a single scoop or a double scoop. How silly. They have nine flavors so I want nine scoops, one of each flavor.

It matters what order the flavors are arranged on the cone. How many different ways could they arrange those nine flavors?

Approach #1: There are 9 possible flavors to go on the bottom. After choosing one of those flavors, there are 8 possible flavors to go next. Then 7 flavors. Then 6 flavors. On so on.

By the fundamental principle, there are 9×8×7×6×5×4×3×2×1 possible arrangements. 9!, which is 362,880.

Approach #2: We want to select 9 items out of 9 items where the order matters. P(9, 9).

From the previous problem (#1187), we know $P(n, r) = \frac{n!}{(n-r)!}$

So $P(9, 9) = \frac{9!}{0!} = 9!$ Note that 0! is defined to equal one.

Just for fun . . .

There are 362,880 arrangements. There are 365 days in most years.

$$\frac{362880}{365} \doteq 994 \text{ years}$$

I could have a differently ordered nine-scoop cone each day for almost ten centuries before I had to repeat.

More fun . . .

If I gained a hundredth of an ounce by eating a nine-scoop cone, then

$$\frac{362880 \text{ cones}}{1} \times \frac{0.01 \text{ ounce}}{\text{cone}} = 3628.8 \text{ ounces}$$

0.01 ounce/cone is a conversion factor.

$$\frac{3628.8 \text{ ounces}}{1} \times \frac{1 \text{ pound}}{16 \text{ ounces}} \doteq 226.8 \text{ pounds}$$

1 pound/16 ounces is another conversion factor.

| 1342–1352 | The Complete Solutions and Answers

1342. Factor $9x^4 + 12x^3y + 16x^6y^6 = x^3(9x + 12y + 16x^3y^6)$
$8w^5 + 16w^{10} = 8w^5(1 + 2w^5)$
$100 + 30z = 10(10 + 3z)$
$49y^4 + 36z^4$ does not factor

1345. Solve $(4x^2 - 3)(3x + 2) = (6x^2 + x)(2x - 5)$

$12x^3 + 8x^2 - 9x - 6 = 12x^3 - 30x^2 + 2x^2 - 5x$

$36x^2 - 4x - 6 = 0$

$18x^2 - 2x - 3 = 0$ Divide by 2 to make things simpler

It doesn't factor. How do I know that? The discriminant, b² – 4ac, is 220, which is not a perfect square.

Use the quadratic formula $x = \dfrac{2 \pm \sqrt{4 - (4)(18)(-3)}}{36}$

$= \dfrac{2 \pm \sqrt{220}}{36}$

$= \dfrac{2 \pm \sqrt{4}\sqrt{55}}{36}$

$= \dfrac{2 \pm 2\sqrt{55}}{36}$

$= \dfrac{2(1 \pm \sqrt{55})}{36}$

$= \dfrac{1 \pm \sqrt{55}}{18}$

1352. Graph $y = 3x^2 + 2x - 7$ from $x = -4$ to $x = 4$.

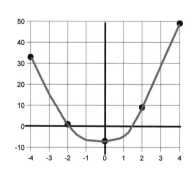

x	y
-4	33
-2	1
0	-7
2	9
4	49

The Complete Solutions and Answers 1359–1366

1359. A system of equations that has exactly one solution (it isn't inconsistent and it isn't dependent) is called independent. Classify each of these systems as 🏵 independent

🏵 inconsistent or

🏵 dependent

$\begin{cases} 3x + 7y = 30 \\ 5x - 7y = 539 \end{cases}$ Adding the two equations we get 8x = 569. Since both variables did not drop out, the equations are independent.

$\begin{cases} -2x + 4y = 44 \\ 2x + 6y = 100 \end{cases}$ Adding the two equations we get 10y = 144. Since both variables did not drop out, the equations are independent.

$\begin{cases} 6x - y = 13 \\ -12x + 2y = 20 \end{cases}$ Multiplying the first equation by 2 and adding the equations, we get 0 = 46. Both variables dropped out. Since 0 = 46 is never true, the equations are inconsistent.

$\begin{cases} 4x + 5y = 8 \\ -12x - 15y = -24 \end{cases}$ Multiplying the first equation by 3 and adding the equations, we get 0 = 0. Both variables dropped out. Since 0 = 0 is always true, the equations are dependent.

1366. The cardinality of a set is the number of elements in the set. For example, the cardinality of {🗶, %, 8, 🏵} is 4.

Suppose g is a function from set A to set B. Suppose g is onto B. Suppose the cardinality of set A is 28. What can you say about the cardinality of set B?

Since g is onto B, every element in B is hit at least once.

Suppose set A is a set of 28 archers, each with one arrow. Imagine set B as a set of targets. Each archer shoots one arrow. (That makes it a function.) There might be 28 targets in the case that each archer hit a different target. There might be only 3 targets (three elements in B) if all 28 archers hit the three targets. But there could not be 29 targets, because, in that case, not all the targets would be hit. We are told that g is onto.

Therefore, the number of targets (the cardinality of B) is ≤ 28.

1380–1394 The Complete Solutions and Answers

1380. Factor $2y^2 + 26y + 84 = 2(y^2 + 13y + 42) = 2(y+6)(y+7)$
Always look for a common factor first

$$x^4 - 81 = (x^2 + 9)(x^2 - 9) = (x^2 + 9)(x+3)(x-3)$$

$20x^2 + 11x - 3$
Look for two things that add to 11x and multiply to $-60x^2$ ($= 20x^2$ times -3)
$= 20x^2 + 15x - 4x - 3$
Factor by grouping
$= 5x(4x+3) - (4x+3)$
$= (4x+3)(5x-1)$

1387. They sang 8 Christmas carols and asked me to choose my favorite, my second favorite, and my third favorite. How many ways could I pick first, second, and third place favorites?

First approach: There are 8 possible carols for me to pick as my favorite. After choosing first place, there are 7 possible carols for second place, and 6 possible for third place.

By the fundamental principle, there are $8 \times 7 \times 6$ ($= 336$) possible ways to pick first, second, and third place.

Second approach: Picking 3 out of 8 where the order matters is P(8, 3).

The formula is $P(n, r) = \dfrac{n!}{(n-r)!}$ $P(8, 3) = \dfrac{8!}{5!} = 8 \times 7 \times 6$

1394. How many different ways can you arrange [L][L][L][P][Q][R][S] in a row?

There are 7 spaces to be filled: ___, ___, ___, ___, ___, ___, ___ .

There are C(7, 3) ways to select where the three L's will go. After that, you have four remaining blank spaces.

There are four choices (P, Q, R, or S) to put in the first of the blank spaces.

There are three choices for the second blank space. Two for the third blank space, and whatever is the remaining tile goes in the last blank space.

By the fundamental principle, [L][L][L][P][Q][R][S] can be arranged in C(7, 3)$\times 4 \times 3 \times 2 \times 1$ ways. $\dfrac{7!}{4! \, 3!} \times 4! = \dfrac{7!}{3!} = 7 \times 6 \times 5 \times 4 = 840$ ways

The Complete Solutions and Answers 1401–1408

1401. $\dfrac{7x+21}{3x^2-17x+10} \cdot \dfrac{x^2-25}{x^2+x-6}$

$= \dfrac{(7x+21)(x^2-25)}{(3x^2-17x+10)(x^2+x-6)}$ Top times top and bottom times bottom

$= \dfrac{7(x+3)(x+5)(x-5)}{(x-5)(3x-2)(x+3)(x-2)}$ Factor top and factor bottom

$= \dfrac{7(x+5)}{(3x-2)(x-2)}$ Cancel like factors

> Here is the factoring of $3x^2-17x+10$
> $3x^2-17x+10$
> $= 3x^2-15x-2x+10$ Splitting $-17x$ into two things that add to $-17x$ and multiply to $+30x^2$
> $= 3x(x-5)-2(x-5)$ Factoring by grouping
> $= (x-5)(3x-2)$

1408. How many possible functions are there for h when we are given h:A → B where A has 47 elements and B has 80 elements?

There are 80 choices for the image of the first element in A.
There are 80 choices for the image of the second element in A.
There are 80 choices for the image of the third element in A.
There are 80 choices for the image of the fourth element in A.
. . .
There are 80 choices for the image of the 47th element in A.

There are 80×80×80×80×80×80×80×80× . . . ×80×80×80 (= 80^{47}) different ways that function h could be created.

In higher math (beyond calculus) the set of all possible functions from set A to set B is written as B^A. Now you know why that notation is used. (If A has 10 elements and B has 8 elements, the set B^A consists of 8^{10} functions.)

227

1422–1429 | The Complete Solutions and Answers

1422. Prove $9 + 13 + 17 + 21 + \ldots + 4n + 5 = n(2n + 7)$ is true for every natural number.

The first step is to prove the n = 1 statement is true. This part of the math induction proof is usually super obvious.

$$n = 1 \Rightarrow 9 \stackrel{?}{=} 1(2(1) + 7) \quad \text{True.}$$

The second step is to assume the n = k statement is true.
Namely, we assume $n = k \Rightarrow 9 + 13 + \ldots + 4k + 5 = k(2k + 7)$

We are allowed to use this fact in trying to prove the n = k+1 statement.
To prove:
$n = k+1 \Rightarrow 9 + 13 + \ldots + 4k + 5 + 4(k+1) + 5 = (k+1)(2(k+1) + 7)$
Let's start with what we are assuming to be true:

$$9 + 13 + \ldots + 4k + 5 = k(2k + 7)$$

What's the next number after $4k + 5$ in the series? It will be $4(k+1) + 5$.

If I add $4(k+1) + 5$ to both sides of the series that I have assumed to be true, I get:

$$9 + 13 + \ldots + 4k + 5 + \mathbf{4(k+1) + 5} = k(2k + 7) + \mathbf{4(k+1) + 5}$$

But this is exactly what I'm trying to show is true.
The details: The right side is $k(2k + 7) + 4(k + 1) + 5$, and I need to show that is equal to $(k + 1)(2(k + 1) + 7)$.

$$k(2k + 7) + 4(k + 1) + 5 \stackrel{?}{=} (k + 1)(2(k + 1) + 7)$$
$$2k^2 + 7k + 4k + 4 + 5 \stackrel{?}{=} (k + 1)(2k + 2 + 7) \quad \text{Simplifying each side}$$
$$2k^2 + 11k + 9 \stackrel{?}{=} 2k^2 + 2k + 7k + 2k + 2 + 7 \quad \text{Simplifying each side}$$
$$2k^2 + 11k + 9 \stackrel{?}{=} 2k^2 + 11k + 9 \quad \text{Done!}$$

1429. Give an example of a series that is both arithmetic and geometric.
Example #1: $5 + 5 + 5 + 5 + 5 + 5 \quad\quad a = 5$ and $d = 0$ and $r = 1$
Example #2: $7 + 7 + 7 + 7 \quad a = 7$ and $d = 0$ and $r = 1$
Example #3: $\pi + \pi + \pi + \pi + \pi + \pi \quad a = \pi$ and $d = 0$ and $r = 1$
Example #4: $\ln 4 + \ln 4 + \ln 4 + \ln 4 + \ln 4 \quad a = \ln 4$ and $d = 0$ and $r = 1$

There may be an infinite number of examples, but they are all pretty much alike.

The minute that you say "d = 0" you force r to be equal to 1.

The Complete Solutions and Answers | 1450–1455

1450. $\dfrac{10x^2 - 30x - 100}{4x^2 - 21x + 5} \cdot \dfrac{4x^2 - 9x + 2}{15x^2 - 60}$

$= \dfrac{(10x^2 - 30x - 100)(4x^2 - 9x + 2)}{(4x^2 - 21x + 5)(15x^2 - 60)}$

$= \dfrac{10(x^2 - 3x - 10)(4x^2 - 9x + 2)}{(4x^2 - 21x + 5)(15)(x^2 - 4)}$ Common factoring first

$= \dfrac{10(x - 5)(x + 2)(x - 2)(4x - 1)}{(x - 5)(4x - 1)(15)(x + 2)(x - 2)}$

$= \dfrac{10}{15} = \dfrac{2}{3}$

Here is the factoring of $4x^2 - 21x + 5$

$4x^2 - 21x + 5$

$= 4x^2 - 20x - x + 5$ Splitting $-21x$ into two things that add to $-21x$ and multiply to $20x^2$ ($= 4x^2$ times $+5$)

$= 4x(x - 5) - (x - 5)$ Factoring by grouping

$= (x - 5)(4x - 1)$

1455. How many possible functions are there for g if g:A → B where A has 100 members and B has 300 members and g is 1-1?

    There are 300 choices for the image of the first element of A.
    There are 299 choices for the image of the second element of A.
    There are 298 choices for the image of the third element of A.
    . . .
    There are 201 choices for the image of the 100th member of A.

    There are 300×299×298× . . . ×202×201 possible functions for g.

    Let's play with the arithmetic:

$300 \times 299 \times 298 \times \ldots \times 202 \times 201 = \dfrac{300 \times 299 \times \ldots \times 201 \times 200 \times 199 \times \ldots \times 1}{200 \times 199 \times \ldots \times 1}$

$ = \dfrac{300!}{200!}$

| 1475 | **The Complete Solutions and Answers** |

1475. Planning a birthday cake is tough. Those star candles cost 2¢ each, give off 3 units of nice smell, and add 5 units of fun. (The fun is lighting the candles and blowing them out.)

The flowers cost 70¢ each, give off 8 units of nice smell, and add 2 units of fun.

We want at least 24 units of nice smell and at least 20 units of fun. How can we do this and minimize the cost?

Our cute little table . . .

	nice smells	fun	cost
x star candles	3	5	2¢
y flowers	8	2	70¢

The constraints
  Smells $3x + 8y \geq 24$
  Fun $5x + 2y \geq 20$
Objective function
  Minimize cost $f(x, y) = 2x + 70y$

Point-plotting $(0, 3)$ and $(8, 0)$ are on $3x + 8y = 24$
  $(4, 0)$ and $(0, 10)$ are on $5x + 2y = 20$

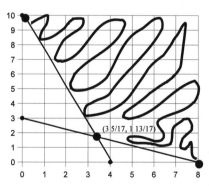

Testing the vertices in $f(x, y) = 2x + 70y$

  $f(0, 10) = 700$
  $f(56/17, 30/17) = 130$
  $f(8, 0) = 16$ ⇐ the cheapest. Use 8 candles and no flowers.

I, your reader, have a question. Suppose the vertex (56/17, 30/17) had given the cheapest alternative. I can't go out and buy 3 5/17 candles and 1 13/17 flowers. What do I do?

Look closely at the graph. There are three points that are in the acceptable (shaded-in) region close to that vertex. I would check (3, 3), (4, 3) and (4, 2) in the objective function.

230

The Complete Solutions and Answers | 1492–1499

1492. If g is a function defined by $g(x) = 5^x$, what is $g^{-1}(x)$ equal to?

Let's do some experimenting.

If $g(3) = 5^3$, then $g^{-1}(5^3) = 3$.
if $g(4) = 5^4$, then $g^{-1}(5^4) = 4$.
if $g(8) = 5^8$, then $g^{-1}(5^8) = 8$.

So in general, $g(x) = 5^x$, then $g^{-1}(5^x) = x$.

What function does this?
$$5^3 \xrightarrow{g^{-1}} 3$$
$$5^4 \xrightarrow{g^{-1}} 4$$
$$5^8 \xrightarrow{g^{-1}} 8$$
$$5^x \xrightarrow{g^{-1}} x$$

$\log_5(x)$ will work.

$$\log_5(5^3) = 3$$
$$\log_5(5^4) = 4$$
$$\log_5(5^8) = 8$$
$$\log_5(5^x) = x$$

1499. I accidentally left the water running in my backyard. In the first minute, it washed away 5 cubic inches of dirt. In the second minute, it washed away 5(0.92) cubic inches. In the third minute, $5(0.92)^2$ cubic inches. Each minute it washed away 92% as much as the previous minute.

After 20 minutes, how many cubic inches of dirt had been washed away?

This is a geometric series. $5 + 5(0.92) + 5(0.92)^2 + \ldots + 5(0.92)^{19}$

$a = 5, r = 0.92, n = 20$

$$s = \frac{a(1-r^n)}{1-r} = \frac{5(1-(0.92)^{20})}{1-0.92} \approx \frac{5(1-0.18869)}{0.08} \doteq 50 \text{ cubic inches}$$

The symbol $\approx$ means "approximately equal to."
The symbol $\doteq$ means "equals after rounding."

The measure of the amount of dirt that was washed away was just a measurement. I measured 5 cubic inches. One significant digit. If I wanted to indicate two significant digits in my answer, I would have written 5.0 × 10. According to the rules for "using significant digits in computing" (see the top of page 122 in *Life of Fred: Advanced Algebra Expanded Edition*) I should have one significant digit in my answer. If I knew that all the numbers in the original problem were exact, then my answer would be more like

$$\frac{5(1-(0.92)^{20})}{1-0.92} = \frac{5(1 - 0.18869332916279655363958709517379)}{0.08}$$

231

The Complete Solutions and Answers

1506. $\dfrac{16w^2 - 16w + 3}{4w^2 - 17w + 15} \div \dfrac{(4w - 1)^6}{16w - 20}$

$= \dfrac{(16w^2 - 16w + 3)(16w - 20)}{(4w^2 - 17w + 15)(4w - 1)^6}$ Invert and multiply

$= \dfrac{(4w - 3)(4w - 1)(4)(4w - 5)}{(4w - 5)(w - 3)(4w - 1)^6}$

$= \dfrac{4(4w - 3)}{(w - 3)(4w - 1)^5}$

> Here is the factoring of $4w^2 - 17w + 15$
> $4w^2 - 17w + 15$
> $= 4w^2 - 5w - 12w + 15$ Two things that add to $-17x$ and that multiply to $+60x^2$
> $= w(4w - 5) - 3(4w - 5)$
> $= (4w - 5)(w - 3)$

Index

adding fractions
- #206. 33
- #277. 34
- #315. 34
- #986. 34

arithmetic sequences and series
- #196. 62
- #36. 62
- #814. 62
- #1429. 64
- #749. 64
- #1130. 68
- #290. 75
- #1258. 76
- #525. 76
- #1297. 76

binomial formula and Pascal's triangle
- #37. 74
- #48. 74
- #68. 74

cardinality of a set
- #1366. 51
- #669. 53

change of base for logs
- #574. 24
- #739. 24
- #345. 24
- #503. 24
- #1117. 26
- #1252. 46
- #126. 75

circle
- #347. 43
- #425. 43
- #770. 43
- #993. 43
- #236. 60

combinations
- #47. 71
- #60. 72
- #1394. 72
- #87. 73

constants of proportionality
- #104. 14
- #1064. 14
- #1102. 14
- #351. 14
- #485. 14
- #50. 14
- #624. 16
- #1007. 26

Cramer's rule
- #166. 38
- #335. 38
- #684. 38
- #1049. 39

cross multiplying
- #1010. 13
- #1028. 13
- #674. 13
- #62. 13
- #1261. 13
- #919. 13
- #431. 13
- #1300. 16

#1222. 20
#569. 26
#1172. 46

degree of a polynomial
#589. 54
#670. 54

determinants
#333. 37
#539. 37
#1237. 38
#769. 38
#874. 38
#1202. 38
#96. 39

distance between two points
#759. 29
#299. 29
#339. 29
#549. 29
#664. 29
#1127. 31
#136. 53

double-intercept form of the line
#383. 30
#719. 30
#1067. 31
#93. 53

ellipse
#1016. 40
#1177. 40
#1207. 40
#834. 40
#540. 40
#914. 40

#1255. 41
#186. 41
#1288. 41
#443. 46
#1294. 46
#231. 53
#236. 60

exponential equations
#1004. 25
#1037. 25
#1107. 25
#694. 25
#527. 25
#413. 25
#859. 25
#305. 26
#609. 31
#266. 39

exponents
#1055. 17
#281. 17
#401. 17
#644. 17
#83. 17
#904. 17

factorial
#86. 71

factoring—common factors
#1342. 32
#554. 32
#293. 32
#1306. 32
#784. 39
#1380. 53

Index

factoring—difference of squares
- #784. 39
- #1380. 53

factoring—easy trinomials
- #296. 32
- #52. 32
- #956. 32
- #1142. 32
- #784. 39

factoring—grouping
- #389. 33
- #500. 33

factoring—harder trinomials
- #325. 33
- #634. 33
- #929. 33
- #784. 39
- #1380. 53

fractional equations
- #959. 34
- #559. 34
- #829. 34
- #1013. 34
- #211. 60

fractional exponents/square roots
- #844. 17
- #1246. 17
- #709. 17
- #107. 17
- #629. 20
- #455. 20
- #824. 26
- #121. 39

functions as ordered pairs
- #66. 49
- #114. 50
- #962. 75
- #654. 75

functions—definition
- #41. 47
- #240. 47
- #337. 48
- #479. 48
- #580. 53
- #445. 60
- #835. 68
- #974. 75
- #300. 75

functions—domain
- #665. 50
- #517. 61
- #995. 68
- #835. 68
- #1315. 69
- #815. 70
- #962. 75
- #45. 75

functions—one-to-one
- #584. 50
- #515. 50
- #446. 50
- #580. 53
- #445. 60
- #995. 68
- #1315. 69
- #977. 71
- #1455. 71

Index

#974. 75
#962. 75
#300. 75

functions—onto
#98. 51
#1366. 51
#580. 53
#669. 53
#445. 60
#995. 68
#300. 75

functions—range
#619. 49
#510. 49
#1240. 51
#962. 75
#45. 75

fundamental principle
#67. 70
#815. 70
#1408. 70
#750. 70
#59. 70
#99. 70
#1455. 71
#977. 71
#1394. 72
#998. 72
#87. 73
#590. 75
#1387. 76

geometric sequences and series
#330. 63
#301. 63
#1429. 64
#963. 64
#749. 64
#1022. 64
#1130. 68
#1499. 69
#1297. 76

graphing by point-plotting
#112. 27
#509. 27
#461. 27
#271. 27
#297. 27
#754. 31
#331. 46
#1313. 61
#1352. 76

graphing inequalities in two
 variables
#395. 45
#809. 45
#614. 45
#875. 45
#473. 53
#774. 68
#191. 75

graphs of a function
#349. 50
#654. 75

Index

hyperbola
- #1162 ... 45
- #804 ... 45
- #724 ... 45
- #960 ... 45
- #1052 ... 45
- #359 ... 53
- #774 ... 68

imaginary numbers
- #1019 ... 19
- #1031 ... 19
- #1122 ... 19
- #849 ... 19
- #497 ... 19
- #251 ... 19
- #171 ... 19
- #659 ... 19
- #285 ... 20
- #313 ... 20
- #1192 ... 20
- #909 ... 20
- #1282 ... 20
- #261 ... 26
- #65 ... 46
- #113 ... 75

inconsistent and dependent equations
- #1359 ... 37
- #794 ... 37
- #971 ... 37

invent a function
- #650 ... 49
- #745 ... 49

inverse function
- #85 ... 51
- #74 ... 51
- #116 ... 51
- #365 ... 52
- #192 ... 52
- #690 ... 53
- #1492 ... 60

language of graphing
- #983 ... 27
- #939 ... 27
- #1147 ... 27
- #714 ... 31
- #38 ... 39

laws of logs
- #1087 ... 23
- #734 ... 23
- #1034 ... 23
- #604 ... 26
- #1112 ... 26
- #953 ... 31
- #181 ... 39
- #110 ... 46
- #56 ... 53

linear programming
- #53 ... 56
- #327 ... 56
- #397 ... 57
- #895 ... 57
- #600 ... 57
- #1167 ... 58
- #1090 ... 58
- #1180 ... 61
- #1228 ... 61

Index

#1025. 69
#1475. 77

logarithms (definition)
#92. 23
#407. 23
#1061. 23
#353. 68

long division of polynomials
#141. 54
#521. 54
#779. 54

math induction proofs
#329. 58
#370. 59
#1422. 61

matrices
#1095. 65
#879. 65
#1264. 66
#1318. 67
#1075. 67
#1217. 68

mean average
#273. 13
#1043. 31
#1234. 68

median average
#35. 13
#201. 13
#146. 13
#101. 13
#819. 20
#1043. 31
#1234. 68

multiplying and dividing fractions
#1401. 34
#1450. 34
#1506. 34
#533. 60

parabola
#131. 44
#470. 44
#799. 44
#84. 44
#915. 44

partial fractions
#318. 56
#390. 56
#450. 56
#564. 56
#839. 56
#975. 56
#691. 60
#1212. 75

permutations
#1187. 71
#1321. 71
#1387. 76

point-slope form of the line
#1276. 30
#1291. 30
#764. 30
#889. 31
#1152. 31
#287. 39
#323. 60
#156. 75

Index

pure quadratics
- #317. 35
- #789. 35
- #968. 35

quadratic equations
- #216. 39
- #309. 46

quadratic formula
- #544. 20
- #161. 35
- #639. 35
- #899. 35
- #989. 35
- #1345. 35

radical equations
- #1001. 19
- #1046. 19
- #1270. 19
- #71. 19
- #168. 19
- #283. 19
- #467. 20
- #1273. 26
- #950. 26
- #321. 31
- #275. 60

rationalizing the denominator
- #1077. 18
- #1249. 18
- #1279. 18
- #1303. 18
- #246. 18
- #303. 18
- #341. 18
- #491. 18
- #599. 18
- #729. 18

scientific notation
- #854. 22
- #980. 22

sigma notation
- #77. 63
- #350. 64
- #749. 64
- #963. 64
- #1130. 68
- #1297. 76

significant digits
- #151. 22
- #176. 22
- #689. 22
- #934. 22

simplifying fractions
- #307. 33
- #699. 33
- #95. 33

slope
- #44. 28
- #579. 28
- #864. 28
- #944. 28
- #965. 28

slope-intercept form of the line
- #1040. 29
- #1137. 29
- #869. 29
- #924. 29

Index

subtracting fractions
- #1072. 34
- #894. 34

systems of equations—by elimination
- #992. 36
- #256. 36
- #279. 36
- #1309. 36
- #221. 53
- #649. 68

systems of equations—by graphing
- #226. 36
- #700. 36

systems of equations—by substitution
- #744. 36
- #679. 36
- #1157. 36
- #419. 60

two-point form of the line
- #343. 30
- #289. 30
- #947. 31
- #291. 60

using constants of proportionality
- #295. 14
- #704. 15
- #311. 15
- #1225. 15
- #884. 16
- #371. 20
- #80. 39

- #245. 60

Venn diagrams
- #1058. 21
- #1082. 21
- #1197. 21
- #1285. 21
- #377. 21
- #437. 21
- #594. 21
- #89. 21

To learn about other books in this series visit

FredGauss.com